EXPLORING COSMIC MELODIES

The Relationship Between Music and Space

Award-Winning Musician &
6x Amazon Best-Selling Author
CARTER FOX

To my Mom and Dad, who always supported my
dreams and passions.

To my brother Brent, for going along for the
adventure.

To all my friends and family who have ever
supported my music and art — thank you.

And to the scientists, engineers, physicists,
astronomers, mathematicians, artists, musicians,
dreamers, and wonderers — to all my fellow *Soulful
Travelers* inspired by the stars and working to make
the world a better place...

This is for all of you.

TABLE OF CONTENTS

INTRODUCTION

Have humans always looked up at the night sky, searching for meaning, solace, and inspiration in its vastness? How has our fascination with the stars shaped the way we create, express, and grow as a society? And what timeless connections exist between the boundless cosmos and the melodies that have echoed through human history?

These are the questions I started asking myself ever since I was a kid, staring up at the night sky through my first telescope, and even more over the past 10 years as I've voyaged on a personal journey of physics and astronomy. I was obsessed as an 8 year old who begged my parents for a closer look at the moon and stars, feeling an overwhelming sense of wonder about what lay beyond. And whose intrigue they supported, perhaps indulged moreso, by gifting that boy a telescope and nurturing that sense of awe.

That wonder only deepened as I learned more about the universe and world around me, imagined more about what's out there and the possibilities of the future, and eventually began translating those feelings into music. I found myself composing pieces and developing entire works that were inspired by the life cycles of stars, the movement of planets, and the grand history of the cosmos, dedicating them to the astronomers, physicists, and dreamers pushing the boundaries of what we know.

But I am not an astronomer. I'm not a scientist or an engineer. I'm simply a person who loves space. I'm a musician, an entrepreneur, and a creative who shares the same fascination with the universe as they do. I may not make discoveries in a lab, but I explore in my own way through sound, emotion, and artistic expression, while continuing to learn new things and stand upon the shoulders of the great thinkers and doers who came before us. And in that, I'm not alone. For generations, across every culture and civilization, humans have looked up at the sky and felt the same pull, the same inspiration, the same need to make sense of it all.

This book is an invitation to a journey that transcends time, space, and identity, exploring the deep connection between the cosmos and music. It's not just about history or science. It's about feeling. About how the stars have shaped not only our understanding of the universe but also our artistic expressions, our creativity, and even our souls. And it's a love letter to the magic that is music the universe has inspired and nurtured us to create.

Together, we'll trace this connection from the ancient world to modern times. We'll uncover how early civilizations infused their music with celestial symbolism, how the concept of the Music of the Spheres imagined the universe itself as a grand symphony, and how harmonic ratios and cosmic proportions laid the foundation for the music we create today.

We'll journey through time, exploring everything from classical compositions steeped in celestial themes to the birth of space music, from the eerie sounds of the cosmos captured by scientists to the futuristic sonic landscapes of science fiction. We'll dive into the world of space-inspired compositions, from the Voyager Golden Record's message to the stars to modern experimental works designed to capture the essence of the cosmos.

But this isn't just a look back, it's also about the future! We'll explore how space continues to inspire new musical frontiers, from electronic and ambient compositions to the ways space travel and exploration shape soundscapes today. And beyond that, we'll reflect on why space resonates so deeply with us as humans and why we see the cosmos as more than just a physical expanse, but as something that speaks to who we are, where we've been, and where we're going.

So, join me on this journey exploring the cosmic symphony that blends history, science, creativity, and emotion. Discover how music and space have been connected since the beginning of our history and how it continues inspiring us. Together, let's explore the melodies of the universe, uncovering how space has shaped our artistic expression and deepening our understanding of our place among the stars.

And so it begins!

CHAPTER 1

THE CELESTIAL SYMPHONY

MUSIC AS AN EXPRESSION OF THE COSMOS

Have you ever looked up at the stars on a clear night and felt like the universe was singing back? Maybe not in a literal sense, but the sheer vastness of the night sky can stir something deep within. Something like wonder, inspiration, love, faith, gratitude, or even a sense of smallness, of being just a tiny note in a much grander symphony.

As a musician, I've taken those feelings and translated them into sound, composing pieces that capture the journey of planets, the birth and death of stars, and the unfolding story of the cosmos. But this isn't just my personal connection. It's something we've been doing our entire existence. For as long as humans have gazed at the heavens, we've turned that cosmic inspiration into music.

Throughout history, the twinkling stars, the celestial dance of planets, and the silent majesty of the cosmos have been more than just objects of scientific study, though their study, itself, is a beautiful artform and science. The stars have been muses, shaping artistic expression, particularly in music for millenia.

This chapter explores that timeless relationship, tracing how ancient civilizations infused their musical traditions with astronomical insight, how harmonic ratios and cosmic proportions became the foundation of musical theory, and how the universe itself has long been seen as the ultimate composer.

Ancient Astronomical Influences on Music

Since the beginning of time, humans have looked up at the night sky in wonder, finding inspiration in the vast, ever-moving cosmos. That inspiration didn't just lead to myths and calendars, it also found its way into music. Across the ancient cultures of the world, people saw the stars and planets as more than distant lights; they were signs, guides, and even instruments in a grand cosmic symphony. Music wasn't just about entertainment; it was a way to tap into the universe's rhythms and align human existence with something much bigger than ourselves.

Egypt: Channeling Celestial Energy Through Sound

Ancient Egyptians were all about the connection between the heavens and daily life. The movement of the stars and planets wasn't just an astronomical curiosity, it dictated religious ceremonies, agricultural cycles, and even the construction of their massive monuments.

Priests, who were basically the rock stars of their time, used music as a way to sync up with the divine. During astronomical events such as eclipses, solstices, phases of the Moon, and appearances of the planets to mark events such as harvests, seasons, and ritual celebrations which were marked by music.

The sistrum, a sacred rattle-like instrument linked to the goddess Hathor, played a big role in rituals. Its jingling sound was believed to drive away evil spirits and bring harmony, literally creating balance between earthly chaos and cosmic order. Harps, lyres, and flutes were tuned to reflect celestial harmony, played in ceremonies that honored gods aligned with specific planets or stars.

To the Egyptians, making music was more than just artistic expression, it was a way to keep the universe in check. Through their music, they bridged the gap between the terrestrial and the divine, weaving together the earthly and the ethereal into a symphony of spiritual resonance and harmonic balance.

Mesopotamia: Math, Music, and the Stars

Across the ancient lands of Mesopotamia, the cradle of civilization, the Babylonians gazed upwards, discerning patterns in the dance of celestial bodies that illuminated the night sky. They had a profound respect for the universe and understanding and honoring it played a large role in their lives.

In their culture, music and astronomy were tightly connected through mathematics. The Babylonians, famous for their detailed star charts and early zodiac systems, saw cosmic patterns in everything including sound. Their musical scales were built on mathematical ratios that mirrored the order they observed in the heavens and the music they used in rituals and celebrations honored the stars and their motion.

Their lyres and harps weren't just for playing simple tunes; they were tuned according to formulas that aligned with planetary movements. Some scholars believe they even composed music based on celestial events, using sound to capture the essence of what they saw in the sky.

To the Babylonians, music was a direct reflection of the order of the universe, a way to communicate with the gods who orchestrated it all, and to make a record of their own existence and place in the universe.

Greece: The Music of the Spheres

The ancient Greeks, known for great strides in art and culture, took things a step further with introducing the idea of the "music of the spheres."

Philosopher-mathematician Pythagoras, of theorem fame, believed that the planets and stars moved according to mathematical equations, producing a kind of cosmic music; one that, while inaudible to human ears, influenced life on Earth. Through such inspiration, he created the first musical scale known as the Pythagroean scale, utilzing musically mathematical concepts known as harmonic ratios and cosmic proportions.

A Brief Exploration of Harmonic Ratios and Cosmic Proportions

Ok, folks. Hold onto your hats a bit and secure your seatbelts. We're about to get a little heavy on the science. As a musician and astronomical enthusiast, this direct relation between the stars and music must be explored. Even the music theory side can get confusing, so I'll keep it brief.

At the heart of this planetary symphony lies the exploration of harmonic ratios and cosmic proportions, a journey that begins with the ancient Greek philosopher Pythagoras and his revolutionary insights into the nature of harmony. In the hushed halls of his academy, Pythagoras postulated that the universe itself was governed by numerical relationships, a divine order that could be understood and expressed through sound. Central to this was his discovery that musical intervals, such as the octave, fifth, and fourth, could be reduced to simple ratios: 2:1, 3:2, and 4:3, respectively. These ratios, Pythagoras believed, were a reflection of the harmony that underpinned the cosmos, a "music of the spheres" that resonated through the heavens, inaudible yet profoundly real.

Applying these principles to music, Pythagoras and his disciples developed a tuning system that sought to encapsulate these celestial harmonies in earthly melodies. The monochord, an instrument used to demonstrate these ratios, became a symbol of the unity between mathematics, music, and the universe. By plucking its single string and adjusting its length to produce specific pitches, akin to stretching a rubber band across a shoebox at different length and plucking it to produce different pitches, they illustrated the connection between numerical precision and auditory beauty. A true tangible bridge between the terrestrial and celestial realms.

The influence of these ideas extended well beyond ancient Greece. In the Middle Ages, the concept of the *quadrivium*, a curriculum of four disciplines — arithmetic, geometry, music, and astronomy — placed music squarely within the realm of cosmic understanding. The belief that music mirrored the proportions and movements of the heavens endured, inspiring medieval scholars and composers to weave celestial symbolism into their works.

The enigmatic allure of the Golden Ratio (approximately 1.618), a mathematical constant found in nature, art, and architecture, also found a home in music. This proportion, often associated with beauty and balance, was revered as a manifestation of cosmic order.

Composers like Béla Bartók and Olivier Messiaen, intrigued by the universality of this ratio, incorporated it into their compositions. Bartók, for instance, structured his pieces using the Fibonacci sequence, a numerical series closely related to the Golden Ratio. Works such as his Music for Strings, Percussion, and Celesta reveal a meticulous alignment with these proportions, creating a sense of organic flow and cosmic symmetry.

Messiaen, similarly, used the Golden Ratio as a framework for rhythm and melody, crafting music that resonates with a profound sense of universal balance.

Greek music theory, heavily influenced by Pythagoras' ideas, was all about mathematical relationships between notes. Instruments like the aulos, a reed instrument, and kithara, a type of lyre, were played with the belief that they could bring listeners into harmony with the universe.

Plato and Aristotle even wrote about music's power to shape human character, seeing it as a force that could align people with cosmic order; something I also wholeheartedly agree with.

India: Sound as the Essence of the Universe

In ancient India, music and the cosmos weren't just connected, they were one and the same.

The concept of *Nada Brahma* ("the universe is sound") is central to Indian philosophy, emphasizing that everything in existence is a form of vibration. The *Samaveda*, one of the oldest Hindu scriptures, describes how hymns were sung to honor the cosmos and maintain balance in the world. These hymns, often performed during astronomical events such as solstices and equinoxes, were believed to maintain cosmic balance and harmony.

Ragas, the intricate melodic frameworks of Indian classical music, evolved from these ancient practices. Each raga is associated with a time of day, season, or mood, reflecting an understanding that music should move in harmony with nature's cycles. This idea still influences Indian music today, reinforcing the belief that sound is a bridge between humans and the greater universe.

The Americas: Music as a Celestial Offering

Indigenous cultures of the Americas also wove astronomy into their musical traditions. The Maya, known for their advanced calendar systems and celestial observations, saw music as an essential part of rituals connected to solstices, equinoxes, and eclipses. Drums, flutes, and rattles were played in ceremonies designed to align human life with the movements of the sun, moon, and stars.

Similarly, the Aztecs used music to communicate with celestial forces, incorporating conch shells and large ceremonial drums to produce sounds that carried through the air like messages to the gods. Their belief in cyclical time—where events repeated in cosmic rhythms—was reflected in their music, reinforcing the idea that sound could influence the balance of the universe.

Native tribes across the northern parts of the Americas have used music as ways to communicate with nature and the universe, preserve and pass down history, and celebrate the events of the cosmos and life. To this day, percussion instruments and vocalisations continue telling stories passed down for generations and reaffirming a connection to the greater universe.

China: The Sound of Harmony

In ancient China, music was deeply tied to cosmology and governance. The belief was that just as the heavens followed precise patterns, so too should human society with music being one of the key ways to maintain harmony between the two.

The imperial court had strict musical regulations, ensuring that certain tones and scales reflected cosmic balance. The guqin, a seven-stringed zither, was especially revered for its meditative qualities and connection to the divine. Playing the guqin wasn't just a musical act; it was a form of self-cultivation, a way to align one's spirit with the greater order of the universe.

African: Music and Spiritualism

Throughout history, African cultures have looked to the stars not only for navigation but as a source of spiritual guidance and inspiration for their music and traditions. The Dogon people of Mali hold one of the most profound cosmic traditions, passing down knowledge of the Sirius star system for centuries through oral history.

Long before modern astronomers confirmed the existence of Sirius B, the Dogon described its orbit, density, and invisibility to the naked eye — knowledge they attribute to the Nommo, amphibious beings from Sirius who visited Earth to share wisdom. This blend of cosmic observation and mythology highlights how the universe has long been seen as a source of both knowledge and mystery.

Further south, the Zulu people of South Africa weave the cosmos into their worldview through star lore, believing the bright star Canopus — known as u-Canzibe — serves as a guiding light for the spirits of the dead. The night sky represents a living connection between the earthly realm and the ancestors, a belief that reflects how music and spirituality intertwine. This cosmic link carries into West African music traditions, where polyrhythms and circular rhythms mirror the cycles of nature and the stars.

In these traditions, drumming patterns often follow the phases of the moon or seasonal changes, turning music into a language that connects humans to the rhythm of the universe.

These cosmic traditions remind us that music is more than sound — it's a universal language that bridges the earthly and the celestial, carrying stories of where we come from and where we're headed. Just like the ancient rhythms echoing through time, music today can help us find our place among the stars.

The Universal Song of the Cosmos

From the pyramids of Egypt to the observatories of the Maya, from the temples of India to the philosopher's schools of Greece, music has always been more than just an art form; it's been a way to connect with the universe. The ancients saw patterns in the stars and translated them into sound, creating music that resonated with the rhythms of existence itself.

Even today, as we study the vibrations of the cosmos through astrophysics and space exploration, we find echoes of these ancient ideas. Whether through the study of planetary frequencies, the discovery of sound waves in space, or our own ongoing fascination with the connection between music and the universe, we continue to listen for that cosmic melody.

The ancients may not have had telescopes or satellites, but they knew something fundamental: the universe is full of music, and we are part of its grand symphony.

Astrology and Music - Celestial Influence on Composition

Throughout history, astrology — the art of interpreting the positions of celestial objects to predict events and human affairs — and music have shared an intricate dance, reflecting humanity's quest to understand, rationalize, and harmonize with the cosmos.

The idea that celestial bodies influence human emotions, behaviors, and creativity found fertile ground in the minds of Renaissance and Baroque composers. During these periods of artistic and scientific awakening, musicians sought to encode the movements of the heavens into their compositions, mirroring the celestial dance of planets and stars through sound.

The Renaissance: A Revival of Cosmic Harmony

The Renaissance, by definition, saw a resurgence of ancient philosophical ideas, including the Pythagorean concept of the "music of the spheres", which, as a reminder, is the belief that planets and stars moved in mathematical harmony, producing an inaudible but ever-present cosmic symphony. Composers of this era believed that musical intervals and structures could reflect the order of the universe, and they embedded astrological symbolism into their works.

Johannes Kepler, the renowned astronomer and mathematician whose discoveries include planetary motion and heliocentrism, took this idea even further in his 1619 treatise *Harmonices Mundi* (*The Harmony of the World*). Kepler theorized that the orbital patterns of planets corresponded to musical intervals, suggesting that the cosmos itself was structured like a grand composition.

His work influenced composers who sought to align their music with these celestial principles, weaving cosmic symbolism into their harmonic choices and melodic structures.

The Baroque Period: Astrology in Composition and Performance

As music became more expressive and intricate during the Baroque period, composers found new ways to integrate astrological ideas into their work. They allowed for emotions to drive sonic choices in new ways, inspired by the heavens above.

Johann Sebastian Bach, though primarily known for his deep understanding of counterpoint and harmony, infused his compositions with mathematical precision that many saw as reflective of cosmic order. His *Well-Tempered Clavier* is often interpreted as an exploration of universal harmony, demonstrating the interplay of order and expression.

In England, composers like John Dowland and William Byrd embraced astrological symbolism in their music. Dowland's melancholic compositions, such as *Lachrimae*, resonated with the astrological association of Saturn, the planet of sorrow and introspection. Similarly, Byrd's sacred choral works were often structured in a way that mirrored cosmic proportions, aligning with the widespread belief that planetary forces influenced human emotion and fate.

Opera and oratorio, two flourishing musical forms of the time, frequently incorporated astrological themes into their narratives. Claudio Monteverdi's *L'Orfeo*, for instance, while primarily a retelling of the Greek myth of Orpheus, subtly reflects the cosmic forces governing love and destiny.

The connection between music and astrology extended beyond composition and into performance, with certain works being scheduled to align with astrological events, believing that celestial timing would enhance their power and resonance.

Even musical instruments themselves were thought to embody cosmic principles. The harpsichord, with its delicate and intricate sound, was often adorned with celestial motifs, symbolizing its connection to the heavens. Its plucked strings were likened to the vibrations of the celestial spheres, creating harmonies that supposedly resonated with the cosmic order. The organ, with its vast tonal range and ethereal presence, was also associated with celestial harmony. Many large pipe organs were built to reflect mathematical proportions found in astrology, reinforcing the belief that music could physically manifest the balance of the cosmos.

Astrology and Music Beyond Europe

The interplay between astrology and music wasn't confined to Europe. In Mughal India, music was deeply intertwined with celestial movements and astrological cycles. Ragas — the complex melodic frameworks in Indian classical music we read about a few sections back — were assigned to specific times of day, seasons, and even astrological events. It was believed that performing a raga at the correct cosmic moment could harmonize the listener with the universe, channeling planetary energies through sound.

In China, traditional tuning systems and musical scales were influenced by Taoist cosmology, which connected musical notes to celestial elements. The ancient Chinese lu scale was structured around astronomical observations, reinforcing the idea that sound and the cosmos were inseparable.

The Modern Echoes of Astrology in Music

Even in modern times, astrology continues to influence musical composition and performance. From Gustav Holst's famous orchestral suite The Planets, where each movement captures the astrological essence of a celestial body, to contemporary artists, such as myself, who create music based on lunar cycles, planetary alignments, and astronomical events, the cosmic and emotional connection remains strong.

Electronic and experimental musicians have also embraced astrological and astronomical themes, using sound frequencies and harmonic structures that resonate with celestial proportions. Some composers even incorporate data from planetary movements into algorithmic compositions, bringing the ancient concept of the "music of the spheres" into the digital age.

Composers like Iannis Xenakis and György Ligeti used mathematical models, including fractals and geometric progressions, to create music that evokes the vastness and complexity of the universe. Xenakis, a trained architect, integrated principles of spatial harmony and mathematical order into his compositions, often drawing parallels between the structures of his music and the patterns of galaxies.

Modern practices in mindfulness and yoga, the emotional stirrings of film scores, the inspired songs of someone who looked up and felt something keep us — mind, body, and soul — connected on Earth and across the universe.

A Cosmic Legacy of Sound

Just as it has always felt throughout my own life, music and the cosmos have always been intertwined, reflecting our fascination with the stars and the endless search for meaning through sound. From the sacred hymns of ancient Egypt to the intricate melodies of Mesopotamia, early civilizations looked to the skies for inspiration, believing that the rhythms of the heavens could shape the music of the earth.

As time went on, this connection only deepened. The Renaissance and Baroque eras brought a renewed focus on astrology's influence, with composers weaving celestial symbolism into their work, each note and harmony an attempt to reflect the grand cosmic order and how we felt in the universe. The music of the spheres wasn't just a theory; it was a guiding force, shaping compositions that mirrored the balance of the universe itself.

At the heart of it all lies the idea that music isn't just an art form, it's a reflection of something greater. From Pythagoras' discoveries about harmonic ratios to the Golden Ratio's impact on composition, music has long been seen as a bridge between the seen and the unseen, the known and the infinite.

Through every era, music has given us a way to translate the mysteries of the cosmos into something we can feel, hear, and understand. It's a tradition that continues today. As we explore the universe through science, sound, and imagination, we carry forward this ancient legacy, harmonizing our existence with the infinite symphony of the stars.

CHAPTER 2

SPACE AS AN INSPIRATION FOR MUSICAL COMPOSITION

Something that has heavily inspired my own music is looking at the stars, learning about astronomy and physics, and further exploring how humans and space are connected to each other. In this inspiration, I am not alone.

The allure of space, with its infinite expanse and celestial wonders, has long stirred the creative imaginations of composers, beckoning them to create music that capture the majesty and mystery of the cosmos. Within the vastness of space, composers find a boundless source of inspiration, where they can explore cosmic themes, evoke ethereal emotions, and beckon listeners on interstellar journeys.

Even personally speaking, space, physics, and learning more about the universe has inspired numerous projects. On my album Epoch, a conceptual album inspired by the history of our universe from the Big Bang up to the moment you are listening to it, I used recordings released by NASA of the orbit of Jupiter as a pad in one of my songs.

On my award-winning album Lost Signals From Outer Space, I created a project intended to emulate the discovery of and eventual communication with an alien race from an initial radio signal burst and attempted to use mathematical elements involving tempo and harmony to represent the universe in music.

Let's explore the multifaceted ways in which space has influenced musical composition, from the classical masterpieces that echo the celestial spheres to the birth of the space music genre and the cutting-edge innovations in sound design and technology that propel us ever closer to the stars.

Cosmic Themes in Classical Music

Throughout much of the period of musical creation that came to be known as the classical era, composers looked to the heavens for inspiration, using the imagery, symbolism, and concepts of space to shape their compositions. This celestial fascination not only reflects humanity's age-old wonder with the cosmos but also serves as a creative conduit for expressing the sublime and the infinite, pushed even further with religious and philosophical tie-ins.

One towering example is Gustav Holst's monumental orchestral suite, *The Planets*. Written between 1914 and 1916, this seven-movement masterpiece captures the astrological and mythological essence of each planet in our solar system (excluding Earth and Pluto, the latter unclassified as a planet at the time).

From the thundering martial rhythms of "Mars, the Bringer of War" to the serene, ethereal beauty of "Venus, the Bringer of Peace," Holst's composition transcends earthly confines, inviting listeners to explore the character and spirit of these celestial bodies. The mysterious undulations of "Neptune, the Mystic," with its ethereal, wordless choir, evoke the vast, unknowable reaches of deep space, leaving audiences in awe.

Similarly, Richard Strauss's iconic tone poem, *Also sprach Zarathustra*, resonates with cosmic grandeur. Its opening fanfare, "Sunrise," has become synonymous with humanity's quest for knowledge and our connection to the universe, immortalized in Stanley Kubrick's *2001: A Space Odyssey*.

Inspired by Friedrich Nietzsche's philosophical work of the same name, Strauss's composition explores themes of creation, evolution, and the infinite cycles of life, mirroring the grandeur and mystery of the cosmos itself. The triumphant brass and organ swell of the fanfare captures the energy of a universe awakening, making it one of the most recognizable and evocative musical tributes to the celestial realm.

Beyond these well-known works, other composers have also drawn inspiration from cosmic themes. Joseph Haydn's The Creation, an oratorio that musically depicts the formation of the universe, offers a celestial narrative rooted in biblical and Miltonic imagery. From the opening chaos of the overture, which represents the formless void before creation, to the triumphant chorus proclaiming "Let there be light," Haydn's work mirrors the awe and majesty of cosmic beginnings.

In the 20th century, Olivier Messiaen explored celestial and spiritual themes in his groundbreaking Des Canyons aux Étoiles (From the Canyons to the Stars), a work inspired by the natural beauty of Utah's Bryce Canyon and the vastness of the heavens. With shimmering orchestral textures and otherworldly harmonies, Messiaen sought to capture the interplay between earthly landscapes and cosmic infinity, bridging the terrestrial with the divine.

Similarly, Alan Hovhaness's Mysterious Mountain embodies a serene yet awe-inspiring reflection on the universe. This symphonic work evokes the majesty of nature and the heavens, with lush, transcendent harmonies that seem to ascend beyond earthly boundaries, inviting listeners into a meditative, celestial space.

As classical composers drew inspiration from the cosmos, their works paved the way for future generations to explore these themes in contemporary contexts. Pieces like John Adams's *Short Ride in a Fast Machine* and Philip Glass's *The Light* continue to echo the connection between music and the vastness of the universe, blending modern techniques with timeless wonder.

These cosmically inspired compositions stand as a testament to the enduring fascination with space and its profound influence on musical expression. By looking to the heavens and far reaches of the universe, composers have created works that not only capture the imagination but also remind us of the limitless possibilities of both the cosmos and human creativity. In their timeless melodies and harmonies, we find echoes of the stars and an invitation to explore the infinite beyond.

The Birth of the Space Music Genre

In the fertile soil of the 1970s, a new genre of music blossomed. It is one that was expressly dedicated to evoking the sensations and imagery of space. Dubbed "space music" or "spacemusic," this genre emerged at the intersection of technological innovation, artistic experimentation, and humanity's growing fascination with the cosmos during the Space Age. It embraced ambient, atmospheric, and electronic sounds to craft sonic landscapes that transported listeners beyond the confines of Earth, creating a bridge between sound and the infinite expanse of the universe.

Pioneers (and personal musical heroes) like Brian Eno, Tangerine Dream, and Jean-Michel Jarre forged sonic odysseys that shimmered with ethereal beauty, using synthesizers, expansive textures, and hypnotic rhythms to evoke the weightlessness and vastness of the cosmos. Eno's groundbreaking album *Ambient 1: Music for Airports* (1978) not only defined the ambient genre but also reflected a profound spatiality in its soundscapes, offering listeners a meditative journey that could just as easily be imagined among the stars as in an airport terminal.

Tangerine Dream, a German electronic music collective, became synonymous with the space music genre, crafting sprawling, otherworldly compositions like *Phaedra* (1974) and *Rubycon* (1975). Their use of sequencers, synthesizers, and layered textures created immersive environments that felt simultaneously futuristic and timeless, evoking the uncharted realms of deep space. Their music not only captivated individual listeners but also found a place in science fiction cinema, where its cosmic qualities perfectly complemented the on-screen exploration of the unknown.

Jean-Michel Jarre's *Oxygène* (1976) further cemented the genre's identity, with its lush electronic textures and soaring melodies capturing the imagination of listeners worldwide. Tracks like "Oxygène IV" became iconic, embodying a sense of wonder and discovery that resonated with humanity's burgeoning ventures into space exploration. Jarre's music, often paired with grand, visually stunning live performances, underscored the expansive and transformative nature of space music, connecting earthly audiences to the cosmos through sound and spectacle.

The genre's reach expanded beyond these pioneers, influencing a generation of artists and composers who sought to explore the relationship between music and the cosmos. Vangelis, for example, brought a cinematic grandeur to space music with works like the *Blade Runner* soundtrack (1982) and *Albedo 0.39* (1976), the latter explicitly inspired by space science and astronomy. The track "Pulstar," with its pulsating rhythms and cosmic ambiance, captures the essence of stars transmitting their energy across the universe.

Space music also found a home in the emerging world of planetarium shows and space-themed media. Composers like Tomita, who reimagined classical works with an electronic twist, brought celestial themes to life in albums like *Snowflakes Are Dancing* (1974), a reinterpretation of Debussy's piano compositions infused with ethereal, otherworldly tones. His music became a staple in planetariums, immersing audiences in a sensory voyage through the stars.

The genre's connection to space exploration itself cannot be overlooked. NASA missions often featured space music-inspired soundtracks, with the Voyager Golden Record, launched in 1977, including musical selections that reflect humanity's connection to the cosmos. While not strictly space music, the inclusion of Bach, Stravinsky, and traditional world music alongside Carl Sagan's carefully curated sounds underscores the idea that music is a universal language capable of spanning the vastness of space.

As technology has advanced, so, too, has the space music genre evolved, incorporating new tools and techniques to push its sonic boundaries. The advent of digital synthesizers, MIDI technology, and computer-based production allowed artists to create even more intricate and immersive soundscapes, keeping the genre vibrant and relevant well into the 21st century.

Contemporary artists like Steve Roach, Biosphere, and Stars of the Lid continue to explore the cosmic themes pioneered by their predecessors, blending ambient, drone, and electronic elements to craft music that resonates with the infinite.

Space music is more than a genre, it's an invitation to dream, to explore, and to imagine humanity's place in the cosmos. Through its shimmering soundscapes and boundless creativity, it continues to transport listeners on journeys to the farthest reaches of space and the deepest realms of their imagination, reminding us of the unending interplay between art, science, and the universe.

Innovations in Sound Design and Technology

Advancements in sound design and technology have propelled the musical representation of space to extraordinary new frontiers. From the advent of early electronic instruments like the theremin, which evoked eerie, otherworldly tones, to the development of sophisticated modern software synthesizers and virtual instruments, the evolution of technology has transformed the way composers translate the mysteries of the cosmos into sound. These tools have enabled artists to sculpt ethereal soundscapes, simulate celestial phenomena, and manipulate audio in unprecedented ways, crafting compositions that transport listeners to the furthest reaches of the universe.

The Moog synthesizer, for instance, revolutionized electronic music in the 1960s, giving composers a means to create entirely new sounds that felt unbound by earthly constraints. Wendy Carlos's Switched-On Bach and her soundtrack work on A Clockwork Orange showcased the vast potential of these instruments to create futuristic, space-adjacent tonalities. Similarly, composers like Vangelis used analog synthesizers to evoke cosmic grandeur in works like the Blade Runner score and Albedo 0.39, where the music itself seemed to pulse with the energy of the stars.

In the digital age, software synthesizers have taken cosmic sound design even further. These upgraded tools allow composers to model complex sonic environments, from the eerie silence of interstellar voids to the chaotic energy of a supernova. Techniques like granular synthesis, sample manipulation, and advanced spatial audio processing add depth and realism, immersing listeners in soundscapes that feel as vast and unknowable as space itself.

The integration of multimedia and immersive technologies has enriched the space music experience, blurring the boundaries between sound, sight, and even touch. Collaborations between composers, visual artists, filmmakers, and technologists have birthed multimedia installations, virtual reality experiences, and live performances that envelop audiences in a sensory symphony. For example, planetarium shows featuring space music, like those scored by Tomita or Jean-Michel Jarre, combine stunning visuals of the cosmos with evocative music, creating a multisensory journey through the universe.

Contemporary artists and composers are pushing these boundaries even further. Virtual reality (VR) and augmented reality (AR) technologies now allow audiences to step inside these cosmic soundscapes, exploring environments where sound and visual elements work in harmony to evoke the wonder of space. Projects like Björk's VR Music Experiences or immersive installations by Brian Eno demonstrate how these technologies can deepen the connection between music and the cosmos.

Sound design has also been influenced by real-world data from space exploration. NASA's sonification projects, which convert data from space telescopes and planetary missions into sound, have added a new dimension to cosmic music. For example, the eerie, haunting tones of the Voyager spacecraft's plasma wave data or the rhythmic pulsations derived from black hole activity provide composers with authentic celestial sounds to incorporate into their works.

In the area of live performance, advancements in audio spatialization and surround sound have transformed the way space music is experienced. Technologies like Dolby Atmos and ambisonics create three-dimensional soundscapes, allowing composers to place listeners at the heart of an interstellar journey.

Visual tracking and projection technology has allowed artists to transport audiences to far off worlds and the depths of the cosmos. Performances by artists like Alva Noto and Ryuichi Sakamoto often utilize these technologies to create immersive environments where sound moves fluidly through space, mirroring the dynamics of celestial motion.

As technology continues to evolve, the possibilities for space-inspired music, and their immersive events, are limitless. Artificial intelligence and machine learning are now being explored as tools to generate music inspired by the patterns of the cosmos, further expanding the creative horizon. With each breakthrough, composers are poised to explore new sonic frontiers, crafting ever more evocative and immersive journeys into the mysteries of the universe.

The fusion of innovation and artistry has opened up boundless opportunities for the musical exploration of space. From the analog synthesizers of the past to the cutting-edge digital tools of the present, and from sonic representations of cosmic data to fully immersive multimedia experiences, space music continues to push the boundaries of human creativity, inviting audiences to embark on profound and transformative musical and experiential voyages through the vast expanse of the cosmos.

CHAPTER 3

THE INFLUENCE OF SPACE EXPLORATION ON POPULAR MUSIC

It's one thing to look up at the stars and be inspired by their grandeur at great distances. But now, humanity has a unique perspective of the truly inspiring aspects of the universe because we've been able to be amongst them. Through manned and unmanned missions, humanity has been able to reach out further into the universe than ever before. The exploration of space has not only expanded humanity's scientific horizons but has also ignited a cultural fascination with the cosmos, leaving an indelible mark on popular culture, particularly in the realm of music.

Let's embark on a celestial journey through the evolution of popular music, tracing its trajectory from the heady days of the Space Age to the cosmic landscapes of space rock, psychedelia, and electronic music. Within this sonic odyssey, we uncover the ways in which space exploration has inspired musicians, shaped musical genres, and captivated the imaginations of artists and audiences alike.

Space Age and the Rise of Sci-Fi Pop Culture

The dawn of the Space Age, marked by groundbreaking achievements like the launch of Sputnik in 1957 and the Apollo 11 moon landing in 1969, ignited a cultural renaissance that echoed across literature, film, television, and music. This era of unprecedented technological advancement and interstellar ambition inspired a collective fascination with the cosmos, blending scientific progress with the imaginative realms of science fiction. Popular music swiftly embraced this sci-fi zeitgeist, with artists channeling the wonders of space into songs that captured humanity's cosmic dreams.

Early rock and roll reflected this fascination, with songs like Jackie Brenston and Ike Turner's *Rocket 88*, often considered one of the first rock and roll records, using the imagery of a rocket-powered car to symbolize the excitement of a new technological era. As the 1960s progressed, instrumental hits like *Telstar* by The Tornados offered a literal sonic tribute to space exploration. Inspired by the launch of the Telstar communications satellite, the song fused futuristic electronic sounds with a catchy melody, becoming the first British single to top the U.S. Billboard Hot 100, cementing its place in musical and cultural history.

The influence of space on pop culture was amplified by the burgeoning science fiction genre in film and television. Movies like *2001: A Space Odyssey* (1968) and *Star Wars* (1977) not only introduced audiences to visually stunning depictions of space but also featured groundbreaking scores that elevated the genre. John Williams' majestic compositions for Star Wars became anthems of intergalactic adventure, blending classical orchestration with the sweeping drama of space opera. Not to mention, for a third time, the epic soundtrack to 2001, which has become iconic space-themed music.

Television shows like *Star Trek* and *Doctor Who* brought weekly doses of cosmic storytelling to audiences, and their influence seeped into music as well. Some blended science heavily into their fiction, while perhaps more relied on fiction than science. But all showed space's influence over our creativity and how fascinated we were and still are with what, and who, is out there.

In music, bands like Jefferson Airplane and Pink Floyd, deeply inspired by the countercultural and exploratory spirit of the era, infused their works with themes of space travel, time, and existential wonder. Pink Floyd's *The Dark Side of the Moon* became an iconic exploration of humanity's relationship with time, space, and mortality, resonating deeply with audiences navigating the vast uncertainties of the Space Age.

Jazz artists like Sun Ra and his Arkestra captured the movements of the planets, the occurrence of astronomical events, and the birth of their own mythology in works that have spanned from the 1950s through today, with members continuing to share the space-inspired stories of aliens and exploration.

The 1970s also saw the rise of glam rock, with artists like David Bowie weaving space-themed narratives into their music and personas. Bowie's *Space Oddity* (1969), released just days before the Apollo 11 moon landing, told the haunting story of fictional astronaut Major Tom, blending melancholy and wonder into an enduring anthem of space exploration. Bowie continued this theme with his alter ego Ziggy Stardust, a cosmic rock star who embodied the intersection of music, science fiction, and the psychedelic aesthetics of the era.

As synthesizers became more accessible, electronic music
further fueled the connection between music and the cosmos.
Albums like Kraftwerk's *Autobahn* (1974) and Jean-Michel
Jarre's *Oxygène* (1976) embraced futuristic sounds, crafting
atmospheric compositions that evoked the vastness of space.
These works laid the groundwork for genres like synthwave
and electronic ambient music, which continue to draw heavily
on space-age aesthetics.

The 1980s and 1990s saw the influence of space expand into
mainstream pop and rock music. Songs like Elton John's
Rocket Man and Europe's *The Final Countdown* became anthems
of interstellar ambition, blending powerful melodies with
evocative lyrics about exploration and discovery. Meanwhile,
science fiction films like *E.T.* and *Blade Runner* inspired
soundtracks that pushed the boundaries of sound design, with
composers like Vangelis crafting atmospheric scores that
mirrored the futuristic worlds depicted on screen.

In the modern era, space and science fiction continue to
inspire artists across genres. Hip-hop artists like Kid Cudi have
embraced space themes, with albums like *Man on the Moon*
exploring introspective journeys through the metaphorical
cosmos. Indie bands like MGMT and Tame Impala weave
psychedelic soundscapes that evoke the mysteries of the
universe, while electronic artists like Floating Points and
Tycho continue to craft ambient works that transport listeners
to otherworldly realms.

The Space Age sparked a creative explosion that forever
altered the cultural landscape, intertwining science fiction and
music in ways that continue to evolve. As humanity sets its
sights on Mars and beyond, the spirit of the Space Age lives on
in the music that dares to dream of the stars, reflecting our
boundless curiosity and the infinite possibilities of the
cosmos.

Space Rock and Psychedelia

The late 1960s and early 1970s birthed a creative revolution in music, giving rise to space rock and psychedelia. These genres sought to break free from the constraints of earthly existence and explore the infinite expanse of the cosmos and emerged as a natural extension of the era's countercultural ethos, blending musical experimentation with a fascination for space, science fiction, and altered states of consciousness.

At the forefront of space rock was Pink Floyd, whose groundbreaking album *The Dark Side of the Moon* (1973) became a defining work of the genre. The record delves into themes of time, human existence, and the vast unknown, with its lush sonic textures and ethereal soundscapes evoking a sense of floating through the cosmos. Earlier works like *A Saucerful of Secrets* and *Meddle* also exemplified the band's ability to weave psychedelic elements into their cosmic explorations, solidifying their status as pioneers of the genre.

Hawkwind, another quintessential space rock band, embraced a more visceral and immersive approach. Albums like In *Search of Space* (1971) and *Space Ritual* (1973) combined driving rhythms, swirling synthesizers, and sci-fi themes to create an auditory experience that felt like an interstellar voyage. Their song *Silver Machine* became an anthem of the space rock movement, capturing the thrill of cosmic exploration with its hypnotic groove and sci-fi lyrics.

David Bowie's Ziggy Stardust persona marked a pivotal moment where glam rock intersected with space rock and psychedelia. *The Rise and Fall of Ziggy Stardust* and *the Spiders from Mars* (1972) introduced audiences to a fictional androgynous rock star from outer space, blending extraterrestrial themes with a message of earthly transcendence. Songs like *Starman* and *Moonage Daydream* became iconic, transcending their era to inspire generations of listeners.

Space rock and psychedelia also thrived in the works of progressive rock bands like Yes and King Crimson. Albums such as Yes's *Close to the Edge* and King Crimson's *In the Court of the Crimson King* blended intricate compositions with surreal, cosmic imagery, creating sonic journeys that felt as expansive as the universe itself.

The instrumentation and production techniques of space rock and psychedelia played a critical role in their ability to transport listeners. Synthesizers, tape loops, and echo effects were employed to create otherworldly textures, while extended instrumental passages and unconventional song structures mirrored the boundless nature of space. Guitarists like David Gilmour (Pink Floyd) and Robert Fripp (King Crimson) used effects pedals and innovative playing techniques to produce sounds that mimicked the vastness and mystery of the cosmos.

Beyond the UK, the cosmic influence of space rock and psychedelia extended to artists like Germany's Tangerine Dream and Kraftwerk, who merged electronic music with interstellar themes, laying the groundwork for ambient and space music genres. Tangerine Dream's *Phaedra* (1974) and Kraftwerk's *Autobahn* (1974) embodied the exploratory spirit of the time, crafting atmospheric compositions that evoked the feeling of drifting through the stars.

Space rock and psychedelia were more than just musical genres; they were cultural movements that provided a sonic sanctuary for a generation seeking escape from the tumultuous realities of the era. Amidst the backdrop of the Space Race, the Vietnam War, and shifting societal norms, these genres offered listeners a gateway to transcendental experiences and interstellar adventures within the boundless expanse of the mind. It united people in space.

Even today, the influence of space rock and psychedelia reverberates through contemporary music. Bands like Tame Impala, MGMT, The Flaming Lips, and Spiritualized draw heavily on the legacy of their predecessors, continuing to push the boundaries of sonic experimentation while exploring themes of space, time, and existence. These modern torchbearers ensure that the spirit of space rock and psychedelia remains alive, inspiring new generations to dream of the stars and embrace the infinite possibilities of the cosmos.

Electronic Music and the Cosmos

The rise of electronic music in the latter half of the 20th century marked a new chapter in humanity's sonic exploration of space. With synthesizers as the primary tools, artists began crafting otherworldly soundscapes that captured the weightlessness, mystery, and vast grandeur of the cosmos. Visionary pioneers like Jean-Michel Jarre, Kraftwerk, and Tangerine Dream were at the forefront of this movement, pushing the boundaries of sound and composition to evoke the infinite expanse of the universe.

Jean-Michel Jarre's groundbreaking albums, such as *Oxygène* (1976) and *Équinoxe* (1978), exemplify the genre's celestial allure. Jarre combined analog synthesizers and lush, atmospheric textures to create music that felt as though it were beamed directly from the stars. His live performances, often featuring elaborate light shows and laser effects, reinforced the cosmic themes, immersing audiences in an audiovisual journey through the cosmos.

Kraftwerk, often regarded as the architects of electronic music, embraced themes of technology and futurism that paralleled humanity's exploration of space. Tracks like *Spacelab* from their album *The Man-Machine* (1978) showcased their ability to blend pulsating electronic rhythms with space-inspired concepts, crafting a sound that mirrored the precision and wonder of space exploration.

Tangerine Dream, with their ethereal, ambient compositions, further deepened the connection between electronic music and the cosmos. Albums like *Rubycon* (1975) and *Stratosfear* (1976) utilized synthesizers, sequencers, and minimalist structures to evoke the sensation of drifting through interstellar voids. Their music became synonymous with the idea of cosmic journeying, influencing not only musicians but also filmmakers who sought to capture the majesty of space on screen.

The influence of space exploration on electronic music extended into the realms of science fiction cinema, where composers like Vangelis used electronic instruments to score iconic films such as *Blade Runner* (1982) and *The Cosmos* series by Carl Sagan. Vangelis's work, particularly his Oscar-winning score for *Chariots of Fire* (1981) and *1492: Conquest of Paradise* (1992), blurred the lines between cinematic storytelling and cosmic soundscapes, inspiring countless musicians to explore similar themes.

In the modern era, electronic dance music (EDM) has embraced and amplified space-inspired themes. Festivals like Tomorrowland and Electric Daisy Carnival often incorporate futuristic visuals, space-themed stages, and immersive light shows, transporting audiences to cosmic dimensions. Producers such as Deadmau5 and Zedd have integrated space-inspired samples and soundscapes into their tracks, creating electronic compositions that echo the awe of the universe.

Tracks like Deadmau5's *Strobe* and Zedd's *Clarity* offer moments of transcendence, where listeners feel as though they are floating weightlessly among the stars.

Beyond just musical performances, the integration of virtual reality (VR) and augmented reality (AR) technologies has transformed electronic music into a multi-sensory experience. Collaborations between technologists and musicians have resulted in immersive experiences where sound and visuals converge, allowing audiences to "travel" through cosmic landscapes while being enveloped in pulsating beats and ethereal melodies.

Space exploration itself has further fueled the creative imaginations of electronic musicians. For instance, Brian Eno's *Apollo: Atmospheres and Soundtracks* (1983), created in collaboration with NASA, was a sonic homage to the Apollo missions. The album's haunting ambient tracks, such as *An Ending (Ascent)* and *Deep Blue Day,* convey the serenity and isolation of space travel, offering listeners an emotional glimpse into the perspective of astronauts gazing back at Earth.

The influence of space exploration continues to shape modern music, spanning multiple genres. Electronic artists like Floating Points and Tycho craft ambient works that transport listeners to otherworldly realms. Hip-hop artists like Kid Cudi have embraced cosmic themes, with *Man on the Moon* (2009) exploring introspective journeys through a metaphorical cosmos. Indie and psychedelic bands like MGMT and Tame Impala weave dreamy soundscapes that evoke the mysteries of the universe.

Even in popular music, the fascination with space endures. Coldplay's *A Sky Full of Stars* is a great example of a mainstream hit that use celestial imagery to convey deep emotions and aspirations.

Meanwhile, technology is pushing space music further than ever before. AI-generated compositions use data from planetary orbits and cosmic radiation to create eerie, mathematical soundscapes. VR and AR technologies allow for fully immersive musical experiences where audiences can "travel" through digital star systems while enveloped in space-inspired sound.

The influence of space exploration on music highlights humanity's enduring fascination with the cosmos and our relentless drive to explore the unknown. From the Space Age to modern EDM festivals, musicians have harnessed electronic soundscapes to express the awe, wonder, and curiosity inspired by the universe. As space technology advances and humans venture further into the stars, it is certain that the cosmos will continue to inspire new genres, sonic landscapes, and immersive experiences that echo the mysteries of the universe.

As humanity continues to push the boundaries of exploration, the relationship between space and sound will only deepen, evolving in ways we have yet to imagine. The stars have always called to us. Through music, we answer.

CHAPTER 4

SPACE SOUNDS AND EXTRATERRESTRIAL MUSIC

The cosmic expanse, with its boundless vistas and enigmatic whispers, beckons us to peer beyond the veil of our terrestrial realm and contemplate the mysteries of the cosmos. Within this infinite tapestry of celestial harmonies, the notion of space sounds and extraterrestrial music emerges as a tantalizing prospect, enticing scientists, musicians, and enthusiasts alike.

What do aliens sound like? What kind of music do they make? Does the Cantina Band's smash hit really hold up in a galaxy far, far away? How loud is a black hole?

It's been a fun creative output by filmmakers, authors, and musicians alike. Even scientists using radio astronomy have begun to sort out the sounds of the very cosmos, itself.

Can we even understand how music would be different, how it affected alien culture differently than our own? Music has been so integral into the development of society. Is that true for civilizations out in space?

Let's continue our odyssey through the celestial symphony, exploring how these alien musings have inspired different terrestrial music.

Capturing and Translating Celestial Sounds

Contrary to the common perception of space as a silent void, the cosmos is alive with a symphony of celestial sounds, from the rhythmic pulses of distant pulsars to the faint whispers of solar winds brushing against magnetic fields. While these phenomena occur at frequencies beyond human hearing, modern technology has enabled scientists to capture and interpret these cosmic melodies, revealing a universe that hums, crackles, and sings with dynamic energy.

Space agencies such as NASA, the European Space Agency (ESA), and various research institutions have employed sophisticated instruments to record these sounds, translating electromagnetic waves, plasma oscillations, and radio emissions into audible frequencies. For example, NASA's Voyager spacecraft, launched in the late 1970s, captured the eerie "songs" of the planets as it journeyed through the solar system. These recordings, like the haunting tones of Jupiter's magnetosphere or the ethereal echoes from Saturn's rings, offer glimpses into the otherworldly symphony of space.

The Chandra X-Ray Observatory has also contributed to this field, converting X-ray emissions from celestial objects into sound. A recent project transformed the data from black holes, galaxy clusters, and nebulae into hauntingly beautiful audio, allowing listeners to "hear" the universe in a new way.

For instance, the Perseus galaxy cluster emits a deep, droning sound; an auditory representation of pressure waves rippling through the cluster's hot gas, akin to the universe's own bass line.

These recordings have not only captivated the scientific community but also inspired artists and musicians to incorporate cosmic sounds into their work. Composer and sound artist Laurie Spiegel famously used Voyager's data to create *Harmonices Mundi,* a piece included in the *Golden Record* aboard the spacecraft. This enduring project, curated by Carl Sagan and his team, sought to represent the sounds of Earth while also imagining the music of the stars. On the title song of my own album Epoch, I used recordings released by NASA that captured the sound of Jupiter's orbit as a pad throughout the entire song.

Translating celestial sounds into audible formats poses unique challenges. The process involves "sonification," where data such as electromagnetic waves or frequency modulations is converted into sound waves that human ears can perceive. Collaborations between scientists and musicians have pioneered innovative methods for this conversion. For example, the "Sounds of Mars" project, using data from NASA's *InSight* lander, transformed seismic vibrations and atmospheric pressure changes into haunting, wind-like audio. Similarly, ESA's *Rosetta* mission captured the "song" of Comet 67P, a fluctuating hum created by the interaction of solar wind and the comet's magnetic field.

These translated sounds have found their way into contemporary music, inspiring artists to merge science and art. Electronic musician Hans Zimmer incorporated space sonification into his score for *Interstellar,* blending synthesized soundscapes with real cosmic data to evoke the grandeur and mystery of space. Similarly, ambient artist Brian Eno has drawn inspiration from space sounds to create immersive tracks that feel like drifting through the stars.

Beyond music, these celestial sounds have enriched multimedia art installations and virtual reality experiences. Projects like *The Space We Live In* use audio recordings from space probes to create environments where participants can "hear" the cosmos while visually exploring star fields and nebulae. Such works bridge the gap between science and art, inviting audiences to experience the universe not just as a vast expanse of stars but as a living, breathing entity with its own unique voice.

The process of capturing and translating these celestial sounds continues to evolve, with advancements in technology promising even deeper auditory insights into the universe. As scientists refine their techniques and expand their explorations, the cosmic symphony grows richer, inviting us to not only listen to the music of the spheres but to embrace the profound connection it fosters between humanity and the stars.

This ongoing collaboration between science and art underscores the universal language of sound, reminding us that even in the vast silence of space, there is music to be heard.

The Golden Record: Music as a Universal Language

A crowning achievement of human ingenuity and aspiration, the famous Golden Record stands as a profound testament to humanity's desire for connection with the cosmos. Launched aboard the Voyager 1 and 2 spacecraft in 1977, this remarkable time capsule carries with it a diverse selection of music, sounds, and images, intended to encapsulate the essence of life on Earth and serve as a greeting to any extraterrestrial intelligence that might encounter it.

At the heart of the Golden Record's mission is the idea that music transcends cultural, linguistic, and even biological barriers, acting as a universal language. Music's ability to convey emotion, tell stories, and evoke shared experiences makes it a powerful tool for bridging divides. With this in mind, Carl Sagan and his team meticulously curated a collection of musical recordings that span the breadth of human civilization, reflecting the diversity and richness of our planet's cultures.

The Golden Record's musical journey opens with Bach's *Brandenburg Concerto No. 2*, a piece that exemplifies the mathematical elegance and timeless beauty of classical music. It also includes Beethoven's *Symphony No. 5* and *Moonlight Sonata*, emblematic of the heights of Western musical tradition. However, the record does not stop at Western classical music. It celebrates the global nature of musical expression.

From the Javanese gamelan's shimmering tones to the spirited rhythms of a Navajo night chant, the Golden Record provides a kaleidoscopic view of Earth's musical heritage. The inclusion of Blind Willie Johnson's haunting blues piece, *Dark Was the Night, Cold Was the Ground*, captures the universality of human longing and spirituality, while Chuck Berry's *Johnny B. Goode* represents the exuberant energy of rock and roll. This eclectic mix highlights not only the technical skill of humanity's composers and performers but also the emotional depth and cultural significance of music across the globe.

The Golden Record also features traditional music from regions often underrepresented in global discourse. For instance, a track of Senegalese percussion showcases the intricate rhythms and communal spirit of African music. The inclusion of an Azerbaijani folk song conveys the lyrical beauty and regional traditions of Central Asia, while Indian classical music, represented by a piece from Ravi Shankar, reflects the spiritual and meditative qualities central to its culture.

Beyond its musical selections, the Golden Record includes natural sounds like thunderstorms, bird calls, and whale songs that underscore the interconnectedness of all life on Earth. These sounds complement the musical tracks, suggesting that humanity's artistic expressions are deeply rooted in the natural world.

The Golden Record's voyage through interstellar space is a poetic gesture, symbolizing humanity's hope to communicate our shared humanity to unknown beings. It reflects the optimism of the Space Age, a belief that our achievements and values could one day reach far beyond our solar system.

While the odds of the record being intercepted by extraterrestrial life are infinitesimally small, its true significance lies in what it represents: a collective effort to define who we are, what we value, and how we express ourselves.

Decades after its launch, the Golden Record remains an iconic cultural artifact, inspiring generations of scientists, artists, and dreamers. Its creation reminds us of the unifying power of music and its ability to transcend time and space. I even have a replica hanging on my wall for inspiration!

Today, as we continue to explore the cosmos, the Golden Record serves as a reminder of our shared humanity; a message in a bottle cast into the cosmic ocean, waiting for a kindred spirit to discover it and join the conversation.

Interstellar Communication and Musical Possibilities

The search for extraterrestrial intelligence (SETI) has not only captivated scientists but also inspired creative thinkers to explore how music might serve as a medium for interstellar communication. The idea of encoding musical information into interstellar transmissions reflects a profound belief in music's ability to transcend cultural, linguistic, and biological barriers. As an art form rooted in patterns, mathematics, and emotional resonance, music offers a unique avenue for bridging the vast gulf of the cosmos, potentially connecting us to other intelligent beings.

Envisioning interstellar musical communication begins with the recognition that certain musical elements such as rhythm, melody, and harmony are based on mathematical relationships that may be universally comprehensible. For example, the harmonic series, which underpins much of Earth's musical theory, arises from natural acoustical properties that could be familiar to extraterrestrial civilizations. By encoding music with deliberate patterns or universal mathematical principles, scientists and musicians hope to create transmissions that convey not only the structure of music but also its emotional essence.

One fascinating example of this concept is the Arecibo message sent in 1974, which included encoded information about humanity and Earth. While not musical in itself, the Arecibo message inspired thinkers to consider how music might be used in similar transmissions. A hypothetical "musical Arecibo" could combine mathematical sequences with melodic progressions, crafting a piece that reflects both human intellect and artistry.

The notion of interstellar music exchanges is equally tantalizing. Imagine a scenario where humanity receives a signal from another civilization that includes an alien composition. How would we interpret their sounds? Could we discern their emotions or cultural context, or would their music challenge our understanding of what music can be? Conversely, what would we choose to send in return? Would it be a Beethoven symphony, a traditional folk song, or a cutting-edge electronic piece? Such exchanges could redefine our conception of music, pushing the boundaries of creativity and fostering a sense of cosmic kinship.

The possibilities extend beyond mere exchange. Collaborative interstellar music projects, while speculative, ignite the imagination. Could advanced technology one day enable real-time musical collaborations across light-years, blending the artistic expressions of disparate civilizations? We can do it across the world with musicians in different cities recording nearly in real-time parts for producers. These ventures might lead to entirely new genres, born from the fusion of human and extraterrestrial creativity.

Additionally, the role of music in shaping humanity's broader interstellar message cannot be understated. Programs like METI (Messaging Extraterrestrial Intelligence) have explored how to craft signals that convey more than just raw data. Music, with its ability to evoke emotion and encapsulate culture, could serve as a cornerstone of such efforts. For instance, a carefully composed interstellar symphony might encapsulate the history of human civilization, combining traditional folk elements with modern electronic techniques to illustrate our journey from primitive beginnings to technological advancement.

The concept also raises profound philosophical questions about the nature of music itself. Is music an inherently human phenomenon, or is it a universal construct that any intelligent species might develop? Would alien civilizations perceive it as an artistic form, a scientific pattern, or something entirely different? These questions challenge us to think deeply about the essence of music and its role in the universe.

As humanity continues to explore the stars, the dream of interstellar musical communication remains a powerful symbol of our curiosity and creativity. Whether through encoded signals, cultural exchanges, or collaborative projects, music offers a hopeful vision of connection; a bridge that spans the cosmic void and unites us in a shared exploration of the mysteries of existence. The idea of creating a universal symphony, one that resonates across light-years, reminds us that even in the vastness of space, we are part of a greater cosmic harmony.

When we look, it's not just what we see that inspires us. Space sounds and extraterrestrial music invite us to think beyond Earthly confines and explore the cosmos through sound. The capture and interpretation of celestial noises reveal a dynamic and vibrant universe, while projects like the Golden Record serve as messages of artistic and cultural expression cast into the cosmic ocean.

The notion of interstellar musical communication challenges us to imagine music as more than an art form. It could truly be a bridge connecting intelligent life across the stars. Whether through scientific discovery or creative interpretation, music continues to serve as humanity's universal language, inviting us to listen, dream, and reach beyond the horizon toward the unknown.

CHAPTER 5

MUSIC IN SPACE

THE SOUNDS OF ASTRONAUTS AND SATELLITES

The boundless expanse of space has long served as a muse for artistic expression, with music emerging as a poignant conduit for exploring the mysteries and wonders of the cosmos. From the earliest days of space exploration to the present era of interstellar voyages, music has played an integral role in shaping the experiences of astronauts and the transmission of space data. After all, humans, no matter where they are, have their connection with music. It's as natural as the rhythmic beating of our hearts.

As a musician, I have always wondered what it was really like to play music in space, to put on a performance in orbit or on the moon, or to have an audience quite literally cheering from around the world. Some of these things haven't happened quite yet, but there are a few amazing people who have actually had the experience to play music in orbit!

This chapter shows us how music touches the human heart and the role of music in space exploration, taking a look at its presence during missions and its influence on astronauts.

Musicians in Space: Apollo 11 and Beyond

The historic Apollo 11 mission, which achieved humanity's first steps on the Moon in 1969, stands as a testament to our audacious spirit of exploration and discovery. Though music itself did not physically accompany Neil Armstrong, Buzz Aldrin, and Michael Collins on their lunar journey, it was undoubtedly present in spirit.

The crew, like their counterparts on subsequent missions, held a deep personal connection to music, a universal thread that offered comfort and inspiration amidst the vast expanse of space. They carried within them the melodies of their homeland and the rhythms of the era, a reminder of the world they were venturing beyond.

Following Apollo 11, music became an integral companion to astronauts, finding its way into the routines and rituals of space travel. By the time of the Skylab missions in the 1970s, astronauts brought cassette players loaded with their favorite music to help combat isolation and maintain morale. The tradition continued to evolve, with musical selections curated for shuttle missions and playlists tailored to individual tastes becoming a staple of long-duration flights.

As human space exploration advanced, so too did music's role in the lives of astronauts. Instruments began accompanying crews into space, allowing them to create music rather than merely listen to it. Canadian astronaut Chris Hadfield famously performed David Bowie's *Space Oddity* aboard the International Space Station (ISS), recording the first music video in space and captivating audiences around the globe.

Hadfield's performance, complete with zero-gravity guitar strumming, epitomized the transformative power of music to bridge the divide between Earth and the cosmos.

Beyond individual performances, music aboard the ISS has sparked collaboration and creativity among astronauts of diverse nationalities. Russian cosmonauts have been known to bring harmonicas and guitars, while NASA astronauts have celebrated holidays with musical performances that blend cultures and traditions. These moments of shared artistry underscore music's capacity to unite people, even in the confines of a space station orbiting 250 miles above Earth.

Music's influence extends beyond the confines of spacecraft. In 2005, British rock band Muse debuted the track *Supermassive Black Hole* with imagery inspired by space exploration. Meanwhile, orchestras have performed live for space missions, such as when the Houston Symphony played during the final shuttle mission, providing a real-time soundtrack to humanity's celestial endeavors.

From the cassette players of the Apollo era to live concerts broadcast from orbit, music has woven itself into the narrative of space exploration. It serves not only as a source of comfort and creativity for astronauts but also as a bridge, connecting them to the world they left behind and the dreams of a future among the stars.

These melodies, strummed and sung in the zero-gravity theaters of space, remind us that even as we journey into the unknown, we carry the essence of humanity — our art, culture, and shared spirit — with us.

Music in Space Missions and Spacecraft

Music has long been an integral part of space missions, offering astronauts more than just comfort; it becomes a vital thread weaving together the human experience in the sterile, high-tech environment of spacecraft. Astronauts, often facing months of isolation aboard vessels like the International Space Station (ISS), meticulously curate personal playlists.

These carefully selected collections of songs evoke cherished memories, inspire motivation during challenging moments, or provide a soothing escape from the monotony and solitude of life in space. Whether it's the rock anthems of Earth, serene classical pieces, or nostalgic tracks tied to personal milestones, these playlists transform the silence of the star-strewn abyss into a backdrop for reflection and connection.

Music isn't just a personal indulgence; it plays a functional role within the framework of space missions. Mission control teams on Earth have cleverly utilized music as a tool for communication and morale-building. Wake-up calls, a cherished NASA tradition, involved selecting songs for astronauts to start their day, often tied to significant events or themes.

During the Apollo 11 mission, for instance, *Fly Me to the Moon* by Frank Sinatra was a playful yet poignant nod to the groundbreaking journey. In more recent missions, families and friends of the crew have also contributed to these selections, adding a deeply personal touch that bridges the distance between Earth and space.

Music has even been used to convey subtle messages. During the Apollo 12 mission, the crew received a wake-up call featuring "I'll Be Seeing You," a song with lyrics that carried a heartfelt promise of reunion. In other instances, mission control used carefully chosen melodies to provide lighthearted morale boosts, such as playing humorous or celebratory songs to mark milestones or ease tensions during high-stakes moments. These gestures reinforce a sense of unity between the astronauts and their teams on Earth, a reminder that they are never truly alone in their endeavors.

Spacecraft themselves have become platforms for musical creativity. As technology advanced, instruments found their way into orbit. Guitars, harmonicas, and even a keyboard have accompanied astronauts aboard the ISS. These instruments have enabled impromptu jam sessions, live performances broadcast to Earth, and even songwriting, transforming the spacecraft into a zero-gravity studio. Again we mention Canadian astronaut Chris Hadfield's rendition of David Bowie's *Space Oddity* recorded aboard the ISS, because it not only went viral but also underscored the profound emotional resonance of music in connecting humanity across the cosmos.

Music extends beyond recreational use to play a role in psychological and cognitive health. Research has shown that music can mitigate stress, enhance mood, and even improve focus, all critical factors for astronauts operating in high-pressure environments. Space agencies have increasingly recognized this, incorporating music into mental health protocols to support the well-being of crews during extended missions.

As humanity looks toward longer missions to the Moon, Mars, and beyond, music will undoubtedly remain a constant companion. It will serve as a thread connecting astronauts to their roots, to each other, and to those they've left behind. Whether used for comfort, communication, or creativity, music in space missions underscores the enduring power of art to enrich the human experience, even in the most extraordinary circumstances.

Music in space weaves together the threads of human creativity, scientific discovery, and cultural expression into a vibrant tapestry of cosmic connection. From the playlists of astronauts orbiting Earth to the integration of melodies within space missions, music transcends language and culture, creating a shared sense of unity that bridges the gap between those among the stars and those on the ground.

As humanity continues its odyssey among the stars, music will remain an enduring ally, serving as a guide, a source of inspiration, and a reminder of the intricate connection between sound, emotion, and our ever-evolving understanding of the cosmos.

CHAPTER 6

SPACE AND MUSIC

A JOURNEY INTO THE UNKNOWN

Imagine looking up on a clear night, not a single piece of light pollution, nothing in your view. You can see stars, planets, the milky backdrop of our galaxy beautifully covering the field of vision. Your soul is filled with this energy, this inspiration, this peaceful notion of being in your place in the universe.

The cosmos, with its infinite expanse and celestial wonders, has always captivated the human spirit, inspiring awe and igniting a sense of curiosity about the mysteries that lie beyond. Music, with its ability to evoke emotions and transcend language, serves as a powerful vessel for exploring the depths of space.

Our spaceship of the imagination now embarks on a mesmerizing voyage into the intertwined realms of space and music, uncovering how music acts as an emotional conduit to the cosmos, hoping to provide understanding behind the symbolism and metaphors found in space-inspired compositions, and illuminating the emergence of space music festivals and concerts that transport audiences to celestial realms.

Music as an Emotional Bridge to the Cosmos

At the crossroads of space and music lies a profound connection that reaches beyond earthly confines. Music, with its boundless ability to evoke emotion, becomes a vessel for exploring the vastness, beauty, and mystery of the cosmos. It offers listeners a glimpse into the uncharted territories of the universe, using sound to convey the awe-inspiring scale of space and our place within it. Through ethereal melodies, pulsating rhythms, and celestial harmonies, music creates a sonic tapestry that not only mirrors the cosmic unknown but invites us to feel it on a deeply emotional level.

Across genres and styles, composers and musicians channel the evocative power of music to capture the essence of space, crafting auditory experiences that transport us to distant worlds. Ambient pioneers like Brian Eno have created soundscapes that evoke the serene vastness of interstellar travel, while composers like Gustav Holst, with *The Planets*, capture the grandeur and mystique of our solar system through sweeping symphonic movements. Contemporary artists such as Sigur Rós and M83 weave celestial imagery into their music, crafting auditory narratives that feel like stargazing through sound.

Electronic music, with its futuristic tones and expansive possibilities, also plays a pivotal role in bridging the gap between music and space. Jean-Michel Jarre's *Oxygène* and *Equinoxe* albums, for example, create immersive sound worlds that evoke the majesty of the cosmos. Similarly, Vangelis' iconic score for *Blade Runner* and his tribute album *Rosetta* draw listeners into meditative states that feel as though they're drifting through the stars.

Live performances have embraced this cosmic connection as well, with bands like Pink Floyd transforming their concerts into interstellar journeys through light, sound, and imagery. Modern electronic festivals often incorporate space-themed visuals, creating an immersive sensory experience that connects audiences with the universe.

In film and television, music amplifies our emotional connection to the cosmos. John Williams' score for *Star Wars* captures the adventure and drama of space exploration, while Hans Zimmer's *Interstellar* uses organ tones and minimalist motifs to evoke the vastness and mystery of space-time.

In each case, music serves as a bridge, transforming the abstract and infinite into something visceral and deeply felt. It allows us to take the unseen, faraway, and gigantic, and turn it into something we can comprehend and be inspired by.

Ultimately, music's ability to evoke the cosmic is a testament to its power as a universal language. It allows us to transcend the limitations of our senses, drawing us into the infinite expanse of the universe while grounding us in the emotions that define our humanity.

Through these melodies, harmonies, and rhythms, we find a unique connection to the stars; a reminder that while we may be tethered to Earth, our imagination and emotions are free to roam the cosmos.

Symbolism and Metaphors in Space-inspired Music

Space, with its boundless expanse and enigmatic beauty, has long captured the imagination of musicians, serving as a profound source of symbolism and metaphors. Artists across genres draw on astronomical phenomena, celestial bodies, and cosmic concepts to craft narratives that delve into existential themes, provoke introspection, and inspire awe. Space becomes more than a physical realm, it transforms into a canvas for exploring the human experience.

The vast, uncharted expanse of the cosmos often serves as a metaphor for the boundless potential of the human spirit. Songs like David Bowie's *"Space Oddity"* use the imagery of a lone astronaut drifting through space to explore themes of isolation, existential uncertainty, and the fragility of human connection.

Similarly, Elton John's *"Rocket Man"* reflects on the tension between ambition and personal sacrifice, using space travel as a poignant metaphor for the emotional distance and longing that accompany extraordinary pursuits.

The shimmering stars, scattered like diamonds across the night sky, frequently symbolize hope and aspiration. In Coldplay's *"A Sky Full of Stars,"* the stars become a metaphor for love and admiration, a light guiding the way through life's darkness. Meanwhile, songs like *"Across the Universe"* by The Beatles use cosmic imagery to evoke timelessness and interconnectedness, with lyrics that float like celestial bodies across the infinite backdrop of the universe.

Space exploration itself often serves as a potent metaphor for humanity's insatiable curiosity and relentless pursuit of knowledge. Bands like Muse delve into this symbolism with albums such as *Black Holes and Revelations*, blending themes of cosmic exploration with reflections on human ambition and existential crises. In classical music, compositions like Holst's famous *The Planets* personify celestial bodies as archetypes, weaving narratives that explore human emotions and the mysteries of the natural world.

The search for extraterrestrial life, another recurring theme, symbolizes humanity's longing for connection and the desire to understand our place in the cosmic order. Artists like Björk in *"Cosmogony"* use space imagery to reflect on creation myths and humanity's relationship with the universe, while the band Sigur Rós often creates music that feels like transmissions from another world, encouraging listeners to imagine and dream beyond Earth's boundaries.

Beyond individual songs, entire albums and genres embrace the metaphorical potential of space. The ambient and electronic genres, pioneered by artists like Brian Eno, craft soundscapes that evoke the stillness and mystery of the cosmos, inviting listeners to journey inward as they contemplate the vast outward expanse. Pink Floyd's *"Eclipse"* from *The Dark Side of the Moon* captures the duality of light and darkness — both literal and metaphorical — using celestial imagery to explore the contrasts inherent in the human condition.

My own project *"Lost Signals From Outer Space"* is a concept album inspired by the radio burst signals we've received from deep space and if the source had been intelligent life looking for other beings in the universe. The album was also symbolic of the calmness, loneliness, and emptiness of space and the pursuits of exploring it as explored by ambient pieces combined with exciting electronic compositions.

Through symbolic representations in music, space becomes a mirror reflecting humanity's deepest questions and desires. It invites listeners to embark on introspective journeys, pondering their place in the universe, the nature of existence, and the enigmatic forces that shape their reality. Whether through the metaphor of a solitary astronaut, the hopeful glimmer of starlight, or the boundless mystery of the void, space-inspired music continues to illuminate the profound connections between the cosmos and the human experience.

CHAPTER 7

THE EMOTIONAL AND INSPIRATIONAL POWER OF SPACE MUSIC

Music, especially music inspired and evoking the grandeur of the cosmos, has an uncanny way of reaching deep into our emotions, sparking thought-provoking reflections, and awakening our imaginations. It draws us into vast, unexplored territories of sound, where each note feels like an invitation to look beyond the confines of our world. To me, it's one of the greatest sources of comfort and inspiration for creating music and other works.

In this chapter, let's explore the profound emotional and inspirational qualities of space music, uncovering how it has the power to transport us to otherworldly realms, stir a sense of wonder, and spark introspection about our place in the universe. From its ability to evoke feelings of awe and isolation to its capacity for stirring personal revelations, space music offers a unique sonic landscape that resonates with the mysteries of the cosmos, the depths of the human experience, and the our very souls.

Transcendence and Emotional Resonance

At its core, space music is more than just a genre, it's a gateway to experiences that stretch beyond the limitations of conventional musical forms. By breaking free from traditional structures, it invites listeners into a sonic realm where the usual boundaries of time, space, and even reality seem to blur. Through expansive textures, ethereal melodies, and ambient soundscapes, space music creates an immersive environment that doesn't just fill the air, but envelops the listener in a transformative emotional experience.

The vastness and otherworldliness of space music evoke a range of powerful emotions, from awe and wonder to introspection and introspective calm. The ethereal quality of the music transports the listener to a space where they can contemplate the infinite; whether that's the mysteries of the cosmos, their own place within it, or the fragile beauty of existence itself. The immersive soundscapes seem to stretch endlessly, inviting reflection on how small and yet significant we are in the grand scheme of the universe.

Music, in this context, becomes a tool for emotional expansion, enabling us to emotionally "travel" through the unseen realms of space, time, and consciousness.

A perfect example of this transcendence is the work of ambient pioneers like Brian Eno, whose *Music for Airports* and *Apollo: Atmospheres and Soundtracks* not only evoke space-like environments but also create emotional depth through minimal, expansive sound. Similarly, early electronic artists like Jean-Michel Jarre and Vangelis, with albums such as *Oxygène* and *Blade Runner* respectively, use sonic textures to transport listeners to vast, imagined landscapes that feel both alien and familiar.

Space music taps into humanity's innate curiosity and desire to understand the unknown. It allows us to confront the vastness of the universe, not with fear, but with a sense of wonder, awe, and the deeply emotional realization that we are part of something much larger than ourselves. This emotional resonance, whether it's a sense of isolation or the connection we feel in the face of the infinite, makes space music a unique experience that speaks to the heart and soul, offering both a journey outward into the universe and inward into the self.

Through these expansive sonic landscapes, space music offers a transformative experience that stays with the listener long after the final note fades.

Inspirational and Reflective Exploration

Space music is much more than a collection of notes and rhythms; it is a powerful catalyst for inspiration and imaginative thought, an invitation to journey beyond the confines of Earth and into realms yet to be discovered. The sweeping, otherworldly sounds transport listeners to distant galaxies and uncharted planets, igniting a sense of wonder and curiosity. Its ethereal textures, paired with expansive soundscapes, evoke the feeling of embarking on a cosmic odyssey encouraging the mind to roam freely and explore the vast unknown with each note.

The infinite possibilities of space, with its boundless reaches and mysteries, serve as a natural wellspring for artistic and intellectual exploration. Space music reflects the vastness of the cosmos, pushing the boundaries of creativity and innovation. Just as space exploration has expanded our understanding of the universe, space music challenges the limits of genre and composition, inviting artists to experiment with new sounds, structures, and technologies. Artists such as Moby, whose electronic compositions for *The End of Everything* reflect his connection with the universe, have each used space music to explore the intersection of sound and imagination. These works encourage listeners to think beyond the immediate and embrace the infinite potential of the unknown.

Space music also provides a profoundly introspective experience. Its expansive, ambient qualities create an atmosphere of serenity, offering a mental space for reflection. The music's subtle, evolving textures mirror the complexities of the inner world, allowing listeners to delve deeply into their own thoughts and emotions.

Whether it's the meditative soundscapes of artists like Steve Roach or the reflective atmospheres of artists such as William Basinski, space music creates a sanctuary for personal growth and self-awareness. The music's sense of vastness and stillness fosters a rare opportunity for introspection, providing a refuge from the noise and rush of daily life.

In this way, space music becomes a tool not just for exploring the cosmos, but for exploring the self. It invites listeners to reflect on their place in the universe, the mysteries that surround us, and the complex emotions that define the human experience. It offers both a call to action for creative exploration and a quiet space for personal discovery, making it an essential soundtrack for anyone seeking to expand their mind, their creativity, and their understanding of the world.

By offering both inspiration and reflection, space music serves as a bridge between the outer and inner landscapes, guiding the listener on a journey that is both outward into the cosmos and inward into the heart and mind.

Science, Philosophy, and the Human Experience

At its core, space music acts as a harmonious bridge, uniting the realms of science, philosophy, and the human experience into a cohesive, transcendent journey. It encapsulates the boundless curiosity and wonder that drive scientific exploration, capturing the awe that accompanies the uncovering of cosmic mysteries.

The vastness of the universe — its galaxies, stars, and enigmatic forces — serves as a profound source of inspiration for composers and musicians, who translate these scientific discoveries into soundscapes and compositions that mirror the grandeur and unknowns of space. The music becomes not only a reflection of these discoveries but also an emotional representation of humanity's collective thirst for knowledge.

In works like *The Planets* by Gustav Holst, where each movement is inspired by the astrological characteristics of our solar system, space music reflects the depth of scientific inquiry. Similarly, the work of Jean-Michel Jarre, particularly his album *Oxygène*, brings the science of space exploration to life through electronic music, layering sounds that evoke the vastness and mystery of the cosmos.

These compositions are not merely musical interpretations of space, but reflections of humanity's scientific aspirations to understand the universe. Through these sonic landscapes, space music transforms the observable phenomena of the universe into an emotional and sensory experience, inviting listeners to feel the wonder of human discovery.

Beyond the realm of science, space music also delves into the philosophical questions that have intrigued thinkers for centuries. It invites contemplation of humanity's place in the universe, pondering the nature of existence and the search for meaning in an infinite cosmos.

Is life a singular anomaly or part of a larger cosmic order? What are the implications of discovering extraterrestrial life?

These questions, which have long been explored by philosophers, are woven into the fabric of space music.

The meditative qualities of artists like Brian Eno and Max Richter, with albums such as *Apollo: Atmospheres and Soundtracks*, create a sonic space for reflection on the fragility of human existence and the vastness of the unknown. The sweeping, ambient soundscapes encourage listeners to reflect on existential themes such as our search for purpose, our fleeting nature, and our desire to understand both the universe and ourselves.

Space music also speaks to the emotional aspects of the human experience. It taps into universal emotions and aspirations like our intrinsic yearning for exploration, connection, and understanding. The unknown beckons us, and space music mirrors that longing. It channels our desire to voyage into the cosmos, to discover and connect with something greater than ourselves.

The soundscapes created by artists like Moby, with his album *Long Ambients 1: Calm. Sleep.*, offer listeners a space to confront their most fundamental desires, from the simple wonder of existing in the universe to the complex desire for deeper connection. Through its evocative melodies and expansive atmospheres, space music gives voice to our dreams of exploration, while simultaneously offering a sanctuary for introspection.

By merging the scientific with the philosophical and the emotional, space music becomes a powerful conduit for exploring the profound questions that define the human condition. It allows us to contemplate the meaning of existence, the mysteries of the cosmos, and our connection to something larger than ourselves. Whether through the awe-inspired exploration of the universe or the intimate reflection on our place within it, space music serves as a sonic journey that integrates the intellect, the heart, and the soul. It offers an invitation to explore not just the stars above, but the depths of the human spirit, reminding us of our endless pursuit to connect, understand, and find meaning in the universe.

Space music wields a remarkable power to evoke emotions, inspire contemplation, and stimulate the imagination. Its ability to transcend conventional musical boundaries and craft immersive sonic experiences enables listeners to experience a sense of transcendence and emotional resonance.

Through its inspirational and reflective qualities, space music sparks curiosity, fosters introspection, and ignites the imagination. It serves as a bridge between scientific inquiry, philosophical pondering, and the human condition, inviting contemplation of cosmic mysteries and universal truths.

Ultimately, space music beckons listeners on a transformative odyssey, connecting them with the awe-inspiring wonders of the universe and the depths of their own inner landscapes, leaving our imagination open to possibilities as vast as the very universe itself.

CHAPTER 8

THE COSMIC CONNECTION

SPACE MUSIC AND HUMAN SPIRITUALITY

We've been on quite the journey together, exploring how space music interplays with our emotions and inspires us, as well as how it has been a part of our culture for thousands of years. Through this interconnectedness, space music has become intricately woven into the fabric of human spirituality, providing a sonic conduit to explore the depths of the cosmos and our existential place within it.

Now, let's further explore the profound connection between space music and human spirituality, unveiling how it acts as a catalyst for transcendence, introspection, and a deep sense of interconnectedness with the universe.

Transcendence and Mystical Experiences

Space music possesses a remarkable ability to lead listeners into realms of transcendence, where the physical world fades away, and what remains is a profound sense of awe, wonder, and a deep connection to the universe.

Through its delicate, ethereal melodies, vast soundscapes, and atmospheric textures, space music creates an auditory landscape that invites the listener to let go of earthly concerns and journey into the unknown.

These compositions are not merely sounds, but gateways to higher states of consciousness, offering a respite from the noise and distractions of everyday life, leading to moments of stillness, introspection, and expansion.

The immersive nature of space music allows listeners to feel as though they are drifting through a boundless cosmic expanse. Artists like Tangerine Dream, with their pioneering work in electronic space music, have mastered the art of enveloping the listener in sound. Their iconic album *Phaedra*, for example, uses synthesizers and lush textures to evoke a feeling of weightlessness, as if listeners are floating through the vastness of space.

Similarly, the ambient compositions of Brian Eno, particularly albums like *Ambient 1: Music for Airports*, offer an auditory space that is both serene and expansive, creating an atmosphere where the listener can lose themselves in the music, finding moments of calm and clarity amidst the cosmic vastness.

As listeners become attuned to these ambient layers, they often report experiencing sensations of timelessness, a blurring of the lines between past, present, and future. This sense of suspended time mirrors the experiences sought in spiritual practices such as meditation, where one enters a state beyond the limitations of everyday life. The music becomes a conduit for an altered state of consciousness, allowing individuals to transcend their individual identities and tap into a sense of unity with the universe.

This experience is often described as a kind of physical and emotional weightlessness where listeners feel liberated from the constraints of their bodies, and their awareness expands outward, embracing the infinite.

The mystical nature of space music also mirrors the experiences described in various spiritual traditions. Just as practitioners in meditation and contemplative rituals seek to dissolve the ego and connect with a higher consciousness or divine presence, space music facilitates a similar journey.

The music creates a sonic space where personal boundaries blur, and the listener can experience a profound sense of interconnectedness with the cosmos. It is as though the individual self dissolves into the expansive universe, and the barriers between the self and the cosmic order fade away.

This is a sensation often likened to a mystical awakening, where the listener feels at one with the universe, connected to something greater than themselves.

The power of space music to evoke such profound experiences is not limited to solo listening sessions. It also finds expression in the immersive environments of live performances and installations. For example, the work of composer and artist Laurie Anderson, known for her multimedia performances, blends space-inspired music with visual elements to create fully immersive experiences. These performances transport the audience into a shared transcendental space, inviting them to explore the boundaries between sound, vision, and the limitless possibilities of the cosmos.

Similarly, the installation work of artists like Ryoji Ikeda, who uses sound and light to create immersive, otherworldly experiences, brings the listener into a space where the mystical and the technological converge.

Ultimately, space music offers an invitation to explore not only the outer reaches of the cosmos but the inner workings of the mind and spirit. It creates a sacred space for reflection, meditation, and connection to the divine. By encouraging listeners to step beyond the confines of their personal identities and tap into a greater consciousness, space music becomes a tool for spiritual exploration, a sonic bridge that leads to mystical experiences.

Through its expansive soundscapes and transcendent qualities, space music provides a pathway to enlightenment, a means of reaching beyond the earthly realm and into the infinite expanses of the universe.

Reflection and Introspection

Space music, with its expansive soundscapes and ambient textures, creates a unique auditory environment that encourages deep reflection and introspection. The vast, open spaces within the music serve as a metaphor for the vastness of the mind and soul, inviting listeners to embark on a journey within themselves.

Its ethereal, often minimalistic qualities act as a gentle guide, allowing the listener to step away from the noise of daily life and enter a space of quiet contemplation. Whether it's the long, drifting drones of Stars of the Lid or the shimmering, atmospheric layers in works by artists like William Basinski, space music creates a sonic landscape that provides the perfect backdrop for inner exploration.

The nature of space music allows for an experience that is both immersive and meditative. Much like the practice of mindfulness or meditation, the music fosters an environment where thoughts can settle, and the listener becomes more attuned to their inner world.

The absence of structured rhythms or melodies in many space music compositions creates a feeling of openness, offering listeners the freedom to explore their own mental and emotional landscapes without the constraints of a traditional musical form. This sense of freedom enables the listener to dive deeply into their personal thoughts, emotions, and even existential musings.

For many, the space music experience becomes a catalyst for moments of clarity, epiphanies, and insights that may otherwise remain hidden in the everyday rush of life. The meditative tones and vast sonic textures open up space for reflection on one's life, purpose, and place within the larger tapestry of existence.

Works like *Music for Airports* by Brian Eno, with its slow-moving, harmonic shifts and atmospheric presence, often inspire listeners to take a step back and reflect on their lives, contemplating both the mundane and the profound. The music creates space for the mind to wander, to question, and to experience moments of self-discovery.

Moreover, space music allows for a profound exploration of life's mysteries, connecting personal introspection with the larger mysteries of the cosmos. In the same way that the music evokes awe and wonder about the universe, it also facilitates an exploration of the deeper questions about existence: *Why are we here? What is our purpose? What lies beyond the limits of our understanding?*

The music, much like the stars it evokes, serves as a vast canvas for these questions, offering listeners the space to explore them without the pressure of finding concrete answers. In this way, space music mirrors our quest for meaning, both on a personal and universal scale.

Artists like Steve Roach, whose albums such as *Structures from Silence* create immersive, ambient environments, have long explored the intersection of space and introspection. Their works invoke a sense of spiritual depth, allowing listeners to reflect on their inner experiences while contemplating the infinite expanse of the universe.

These soundscapes act as both a mirror and a guide, leading the listener through their own emotional and spiritual landscapes while simultaneously connecting them to the larger mysteries of existence. The introspective qualities of space music often result in a feeling of unity between the individual and the cosmos, blurring the lines between the self and the infinite.

Space music also provides a platform for emotional catharsis, helping listeners to process complex emotions or unresolved feelings. The openness of the sound allows for a deep emotional release, whether that manifests in a sense of peace, reconciliation, or even vulnerability. The expansive nature of the music can amplify emotional states, offering a safe space for the listener to confront difficult emotions or to simply experience the fullness of their own being. In this way, space music becomes a tool for emotional healing, allowing listeners to connect with the deeper, sometimes hidden aspects of themselves.

The reflective power of space music is not just limited to the individual; it has the potential to foster collective introspection as well. In immersive environments like planetarium shows or large-scale live performances, space music brings people together in shared experiences of contemplation and reflection.

These events often become communal spaces for introspection, where individuals are united by the sound and the cosmic themes it evokes. Whether in a solitary listening session or as part of a collective experience, space music facilitates moments of deep reflection, offering listeners the opportunity to connect with themselves, each other, and the universe in ways that words alone cannot convey.

In essence, space music serves as a bridge between the mind, the spirit, and the cosmos. It opens a doorway to introspection, inviting listeners to explore their inner worlds while contemplating their place in the vast universe. Through its expansive, meditative soundscapes, space music fosters personal growth, spiritual connection, and an ongoing journey of self-discovery.

It becomes not just a soundtrack to the mysteries of the cosmos, but a mirror to the mysteries of the human experience itself.

Interconnectedness and Cosmic Unity

Space music invites us to explore the vastness of the universe, offering listeners a unique opportunity to contemplate the interconnectedness of all things. Through its celestial melodies and expansive soundscapes, it provides a sonic reflection of the intricate, often unfathomable web of relationships that bind the cosmos together.

In many ways, space music encapsulates the fundamental idea that everything in existence, whether a distant star, a swirling nebula, or the human experience, is part of a larger, interconnected whole. This awareness fosters a sense of cosmic unity, where listeners become acutely aware of their place in the grand scheme of things, often evoking a deep sense of humility and reverence for the universe and the forces that govern it.

The expansive, ambient qualities of space music prompt reflection on the infinite scale of the cosmos, which in turn highlights the transient, yet meaningful, nature of human existence. Composers like Vangelis, known for his iconic *Blade Runner* soundtrack and *Mythodea*, create sonic landscapes that evoke the vastness of space while simultaneously conveying the profound beauty and fragility of life. The music's sweeping, ethereal qualities mirror the way the cosmos stretches beyond comprehension: vast, silent, and awe-inspiring, yet intricately woven together in ways we're only beginning to understand.

This interplay between the infinitesimal and the infinite encourages listeners to consider the delicate balance between the individual and the universe. A reminder that despite the overwhelming size of the cosmos, every star, every planet, and every life form plays a crucial part in its unfolding story.

Space music also serves as a reminder of the interconnectedness of all living things. Just as the celestial bodies are bound by the forces of gravity and physics, we, too, are bound to each other and to the universe. Music, as a universal language, taps into this deep sense of connection, offering a space where listeners of all backgrounds can come together and feel united in their shared experience of existence.

Artists like Brian Eno, with his ambient works such as *Apollo: Atmospheres and Soundtracks*, create music that speaks to this interconnectedness by blending electronic textures with organic sounds, blurring the line between the natural and the artificial. The result is a sonic environment that feels both familiar and otherworldly, creating a connection between the listener's internal world and the expansive, cosmic one.

Moreover, space music transcends the boundaries of culture, language, and religion, acting as a unifying force that can bring people together across vast divides. Just as the stars and the planets exist beyond the confines of our earthly differences, space music offers a spiritual resonance that speaks directly to the soul.

The music's universal themes of exploration, discovery, and awe invite listeners from all walks of life to reflect on their place in the universe and share in the collective human experience of wonder. Whether listening to the expansive soundscapes of *Music for Airports* by Brian Eno or the otherworldly compositions of Jonn Serrie, listeners from diverse cultural and spiritual traditions are able to find solace in the music, as it opens a space for contemplation that is not bound by earthly distinctions.

The sound of space music evokes a sense of belonging to something far larger than ourselves, and in doing so, it offers a profound sense of comfort and connection.

For example, the music of *The Planets* by Gustav Holst, with its sweeping orchestral arrangements, reflects the cosmic drama of the planets and their movements, creating a sense of awe and reverence for the larger universe while also making an emotional connection with the listener. Each piece feels like a meditation on the individual forces that shape the cosmos — gravity, motion, and time — while also expressing the underlying unity of these forces as part of a greater cosmic order.

Space music's ability to transcend cultural and religious divides is particularly significant in a world where differences often separate us. In this context, space music can be seen as a form of spiritual communion, offering a space where individuals can find peace and connection regardless of their background.

For instance, in more immersive experiences like planetarium shows or electronic music festivals, audiences from all walks of life come together, united by the immersive nature of the music and visuals. The shared experience of being enveloped in a soundscape inspired by the cosmos fosters a sense of collective unity, allowing listeners to feel as though they are part of something much larger than themselves.

At its heart, space music is not just about the exploration of space, it's about exploring our relationship to the universe, our connection to each other, and the deeper truths that bind us all. As we listen to the expansive melodies and ethereal harmonies, we are reminded that we are not isolated beings but are part of an intricate, interwoven cosmic fabric. The music, like the universe itself, holds space for reflection, connection, and the exploration of what it means to be part of a greater whole. Through its universal language, space music provides a sanctuary for all who seek to understand their place in the vast, interconnected expanse of existence, offering a sense of unity and belonging that transcends the boundaries of time, space, and culture.

The connection between space music and human spirituality is both profound and symbiotic. Space music acts as a bridge to transcendence, offering us more than just a sonic experience by inviting us to connect with the vastness of the cosmos and to contemplate our place within it. Its ethereal melodies and expansive soundscapes create portals to cosmic wonder, allowing us to feel a deep sense of unity with the universe.

This music becomes a powerful vehicle for introspection and personal growth, providing a space where we can explore the mysteries of existence and reflect on the deeper questions of life.

Through its evocative power, space music nurtures a sense of interconnectedness, reminding us that we are part of a larger cosmic tapestry. Whether through ambient, electronic, or orchestral compositions, space music has the ability to inspire a shared spiritual journey, fostering a sense of harmony that unites us, regardless of background or belief.

In the boundless expanse of space music, the human spirit finds resonance with the mysteries and beauty of the cosmos, encouraging us to embark on a timeless journey of spiritual exploration and discovery.

CHAPTER 9

THE FUTURE OF SPACE MUSIC

Just like the universe itself, we've been on a grand journey of expansion, expanding our minds and becoming more in tune with space has influenced music we create and how the music we create influences the world around us. Space music is an earnest reflection of mankind's obsession with the stars. As humanity continues its journey into the depths of space exploration and embraces ever-evolving technological advancements, the future of space music is poised to embark on thrilling new frontiers.

So before our grand adventure draws to a close, we need to look towards the future to see what's next for our obsession. We will not only explore the potential directions but also look into the innovative technologies and interdisciplinary collaborations that are likely to shape the future of space-inspired musical composition.

Technological Advancements and Sound Design

Technological advancements are reshaping the very foundation of music production and sound design, giving rise to groundbreaking developments in space music. As technology evolves at an unprecedented pace, the future of space music is increasingly intertwined with innovations in virtual reality (VR), augmented reality (AR), and artificial intelligence (AI), all of which open up new realms of sensory experience and creative possibilities.

One of the most exciting aspects of this technological evolution is the growing potential for immersive audiovisual experiences. As VR and AR technologies continue to advance, they offer the possibility of completely enveloping audiences in the cosmos. Imagine stepping into a concert hall where the boundaries between reality and imagination blur where you are not just listening to space music, but experiencing it firsthand.

With immersive soundscapes and 360-degree visuals, audiences could find themselves floating through celestial landscapes, traversing distant galaxies, or flying alongside shooting stars, all while being bathed in the ethereal melodies of space music.

This type of experience pushes the limits of what we traditionally think of as a concert, offering transformative journeys that allow listeners to transcend their physical surroundings and embark on a deeper connection with the cosmos.

Beyond VR and AR, the integration of artificial intelligence (AI) into the creation of space music holds transformative potential. AI algorithms, capable of analyzing vast datasets that incorporate everything from space-related audio signals to astronomical phenomena and musical patterns, can work as collaborators in the creative process. AI-driven compositions could be inspired by the sounds of distant planets, the frequencies of electromagnetic waves, or even the rhythms of celestial bodies.

These unique sonic explorations, generated by AI, could introduce entirely new genres of space music, ones that are more deeply connected to the scientific data that shapes our understanding of the universe. For instance, imagine an AI system that translates the sound of a pulsar or the movements of a comet into a musical composition, creating a piece that resonates not just with human emotions but also with the very fabric of the cosmos.

Furthermore, AI's role in sound design extends beyond composition. It can enhance the creation of complex soundscapes that blend organic and synthetic elements, crafting immersive environments that evoke the mystery and wonder of space.

With the ability to analyze and adapt to listeners' reactions in real-time, AI could personalize the experience, tailoring the music to resonate with individual listeners in unique ways. This level of interactivity could create an entirely new dimension of connection between the audience, the music, and the cosmos.

Together, the convergence of VR, AR, and AI technologies is driving a new era of space music that pushes the boundaries of creativity and sensory experience. With these tools, musicians, composers, and sound designers have the opportunity to break free from traditional methods of music creation, offering audiences a richer, more immersive journey into the cosmos. One where the line between sound, sight, and feeling fades, and space music becomes an all-encompassing, transformative experience. As technological advancements continue to evolve, the future of space music promises to be a dynamic fusion of art, science, and innovation that will forever change the way we listen to the universe.

Interdisciplinary Collaborations

The future of space music is poised to be a rich tapestry woven from the threads of interdisciplinary collaboration, even more than it already is. As the boundaries between art, science, and technology continue to dissolve, musicians, scientists, visual artists, and technologists are coming together in exciting new ways to explore the cosmos through sound. These collaborations are opening doors to unprecedented creative possibilities, transforming the way we experience space and its mysteries.

One of the most groundbreaking aspects of these partnerships is the collaboration between musicians and scientists. Astrophysicists, astronomers, and space agencies are now offering musicians access to real-time space data, allowing them to interpret cosmic phenomena through musical compositions. Imagine a composer using data from a distant galaxy, the rhythmic pulses of a neutron star, or the frequencies of solar flares to craft a piece of music that resonates with the very essence of the universe.

These collaborations blur the lines between scientific discovery and artistic expression, turning raw data into immersive sound experiences that transport listeners to the farthest reaches of space.

The collaboration between NASA and composer Philip Glass for the soundtrack of *The Fate of the Cosmos* serves as an example of how real scientific data can inspire compositions that capture the grandeur and mystery of the cosmos. My song *M87* was inspired by the publication of the first-ever picture of a blackhole that was captured using the Event

Horizon Telescope and dedicated to the team of scientists and astronomers who made it possible, using synthesizers and effects to mimic radio signals beamed through space translating the imagery.

These interdisciplinary collaborations are also giving rise to multimedia installations and live performances that combine space music with cutting-edge technologies. Through the use of augmented reality (AR), virtual reality (VR), and projection mapping, audiences can experience an entirely new dimension of music. Imagine walking through an art installation where the sounds of distant planets and galaxies are accompanied by dynamic visualizations of space, such as swirling nebulae or the movement of celestial bodies.

These immersive environments help audiences feel as though they are part of the cosmic journey, heightening their emotional connection to the music and the mysteries it represents.

The work of artists like Ryoji Ikeda, whose data.scan installations combine sound and digital visuals to explore the limits of perception, illustrates the potential of these kinds of immersive, cross-disciplinary experiences.

Extending beyond musical performances, the fusion of space music with other art forms like dance, theater, and film is creating dynamic new opportunities for storytelling. Collaborations with filmmakers, choreographers, and stage directors are transforming space music into a multisensory narrative experience.

For example, space-inspired music can serve as the emotional backbone for a theatrical performance, where the music guides the movement of dancers or actors, adding layers of meaning to the narrative. In dance, space music can evoke a sense of weightlessness or gravity, allowing movement to mirror the cosmic phenomena the music represents. The film *Interstellar*, with its Hans Zimmer-composed score, is a prime example of how music can elevate a cinematic experience, using space music to underscore themes of love, time, and humanity's place in the universe.

These collaborations also extend to live performances, where musicians work with visual artists to create synesthetic experiences where sound and sight become intertwined to evoke a deeper emotional response. Imagine a live concert where the music is accompanied by projections that mirror the sonic landscape, creating a fully immersive experience for the audience.

The works of artists like Brian Eno, who has long experimented with ambient music and visual art, show the power of combining sound with visual elements to create an environment that is both meditative and dynamic. As technologies like 3D mapping and holography become more accessible, the potential for future space music performances to transform into interactive, multisensory journeys grows exponentially.

In these interdisciplinary collaborations, space music transcends its traditional boundaries, becoming a vehicle for exploring not just the cosmos, but the very nature of creativity itself. The blending of scientific insight, technological innovation, and artistic vision allows space music to evolve into a form of expression that speaks to the mind, the heart, and the spirit.

As these collaborations continue to flourish, they will undoubtedly redefine our relationship with space, art, and the infinite possibilities that lie ahead.

Cultural and Global Perspectives

As humanity continues its journey into space, the exploration of the cosmos is no longer confined to the borders of nations or cultures. The future of space music will undoubtedly reflect this global and multicultural evolution, embracing the rich diversity of artistic expression from all corners of the world.

Musicians, inspired by their unique cultural perspectives, will bring their personal experiences and interpretations of space to the forefront, weaving them into compositions that transcend geographical, linguistic, and cultural boundaries.

Artists from across the globe will draw from their respective traditions, instruments, and musical languages to create space-inspired compositions that are as diverse as the cultures from which they originate.

For example, a composer from India might infuse traditional ragas and the tonal system of Indian classical music into a space composition, interpreting the cosmos through the lens of ancient spiritual philosophies and cosmic concepts.

Similarly, a musician from West Africa might incorporate rhythmic patterns and ceremonial chants that reflect a different view of the universe, blending the vibrancy and spirituality of African musical traditions with the exploratory themes of space music.

These diverse cultural influences will enrich the sonic landscape of space music, creating a more inclusive and expansive narrative about humanity's place in the cosmos.

This fusion of cultural elements will foster deeper cross-cultural dialogues and collaborations. As music from different corners of the globe begins to converge, the collective imagination of humanity will be expressed through a shared love for space and exploration. Collaborative projects between musicians from different cultural backgrounds will allow for unique blending of sounds, creating innovative cross-genre works that reflect the universality of human curiosity and creativity.

A prime example of this is the collaboration between Japanese composer Isao Tomita and various global influences in his work, where he integrated traditional Eastern sounds with cutting-edge synthesizer technology to create space music that transcended both cultural and technological boundaries.

The rapid democratization of music production and distribution through digital platforms is fundamentally transforming the way artists create and share their work. With tools like DAWs (digital audio workstations), online collaboration platforms, and social media, artists from remote areas and marginalized communities now have access to the same resources as those in major urban centers.

This accessibility enables space music to flourish in ways previously unimaginable, as artists from all walks of life can now share their space-inspired compositions with a global audience.

The rise of platforms like SoundCloud, Bandcamp, and YouTube has already given birth to a more diverse and inclusive music scene, where independent musicians can reach listeners across continents without the need for major label support.

As a result, space music will continue to evolve into new subgenres, hybrid forms, and experimental sonic explorations. Artists may merge traditional instruments with cutting-edge digital technology, blending acoustics and electronics to create unique soundscapes that reflect the limitless possibilities of space.

The emergence of genres like space-jazz, ambient-ethno fusion, and cosmic techno exemplifies the creative potential that lies in these diverse musical expressions. For instance, the fusion of African rhythms and jazz with ambient space music, as demonstrated in the work of musician and composer Sun Ra, can produce an exhilarating and introspective listening experience that bridges cultural gaps while still celebrating the vastness of the universe.

This interconnectedness will also drive innovation in live performances and installations, where artists from different cultural backgrounds collaborate to create immersive experiences that showcase the global nature of space exploration. Imagine an international space music festival where artists from around the world come together to perform in a virtual reality setting that transports the audience to the far reaches of the galaxy.

Visual artists, sound designers, and musicians from various cultural traditions could collaborate to create a multimedia experience that reflects the collective human desire to understand the cosmos. This fusion of art and technology would not only amplify the emotional resonance of space music but also emphasize the universal nature of the human spirit's curiosity and creativity.

In essence, the future of space music will be ever-expanding like the universe itself, woven by a vibrant tapestry woven from the threads of cultural diversity, technological innovation, and global connectivity. As humanity looks to the stars, space music will serve as a reflection of our shared exploration of the unknown and our collective aspirations. By embracing multiple cultural perspectives and artistic expressions, the sonic landscape of space music will be enriched, creating new pathways for understanding and connecting with the cosmos.

Space Music Festivals and Concerts

In recent years, the intersection of space, music, and technology has given rise to a new wave of immersive, otherworldly experiences, transforming festivals and concerts into cosmic journeys. These events, designed to push the boundaries of both musical and visual expression, immerse attendees in multi-sensory odysseys that transcend the traditional concert setting.

Space music festivals and concerts blend technology, art, and sound to forge a new kind of connection between performers and audiences, offering a chance to explore the vast mysteries of the universe through music.

Planetarium concerts, with their stunning projections of celestial bodies and interstellar phenomena, have become a popular venue for space-themed performances. These events often feature live orchestras or ambient composers performing alongside synchronized visualizations of the stars, galaxies, and nebulas. One shining example is *Pink Floyd's "The Dark Side of the Moon"* planetarium experience, where the album's iconic soundscapes are paired with mind-bending visuals that reflect the album's exploration of time, existence, and the human condition.

This fusion of sound and visuals invites listeners to truly *feel* the music as they are enveloped by the cosmic imagery around them.

Outdoor festivals, often held in locations far from city lights, take the cosmic experience to the next level by setting the stage beneath a real starry sky. These festivals, such as the Cosmic Convergence Festival or Earth Frequency Festival in Australia, combine ambient, electronic, and psychedelic music with interactive lighting and projection mapping.

Artists like Shpongle and Ott, known for their intricate, spacey compositions, use these festivals as a platform to showcase their otherworldly sounds and mesmerizing visuals, creating an environment where the audience is taken on an auditory and visual journey through the cosmos. The natural surroundings, coupled with ethereal soundscapes, provide a powerful sense of connection to the universe and the earth beneath our feet.

In the realm of more experimental and progressive genres, festivals like *The Sonic Bloom Festival* and *Symbiosis Gathering* embrace space-themed performances that blur the line between music, art, and technology. These festivals often feature immersive art installations, interactive sound sculptures, and virtual reality experiences that extend the concept of space beyond physical boundaries.

Progressive rock bands like *Yes, King Crimson*, and *Pink Floyd* have long been associated with the concept of space music, and their live shows offer an immersive experience that combines elaborate light shows with expansive, otherworldly soundscapes.

These performances often evoke the feeling of journeying through space, with vast, swirling instrumental passages that mirror the exploration of unknown realms. The live renditions of *"Close to the Edge"* by Yes or *"Echoes"* by Pink Floyd, for example, unfold like sonic explorations of distant galaxies, tapping into the infinite possibilities of sound and space.

The rise of space-themed festivals has also brought together artists who may not traditionally be associated with the genre, but whose music evokes the deep wonder of the cosmos.

Electronic artists like Tycho and Bonobo create lush, atmospheric soundscapes that resonate with listeners on a visceral level, often incorporating ambient, space-like textures into their live shows. Their performances, often set against vast, immersive projections of the natural world or cosmic imagery, offer audiences a chance to transcend the boundaries of time and space, experiencing the beauty and mystery of the universe through sound and light.

One of the most exciting recent developments in space-themed concerts is Las Vegas' *Sphere*, an innovative entertainment venue that takes the concept of immersive performance to an entirely new level. The *Sphere* is a state-of-the-art venue with a massive, 360-degree, high-definition spherical screen that wraps around the audience, creating a fully immersive environment.

The venue's exceptional technology allows for dynamic, interactive visuals and spatial sound design, transforming concerts into fully realized cosmic journeys. It's equipped with a custom-designed sound system that makes use of spatial audio, meaning that sound can be placed anywhere in the venue, enhancing the feeling of being enveloped by the music.

Artists like *Phish, Dead & Company, U2,* and *Billie Eilish* have already utilized the *Sphere*'s cutting-edge technology to create experiences that feel like space odysseys. The concerts are designed to make audiences feel as though they are floating through space, surrounded by stars, planets, and nebulae.

This level of technological innovation makes the *Sphere* an ideal venue for space-themed performances. Whether through synchronized visuals of galaxies, nebulae, or abstract representations of space, the Sphere is redefining the live concert experience, allowing music to transcend mere performance and transport attendees into the very fabric of the cosmos.

As technology continues to evolve, space music festivals and concerts will likely become even more immersive, incorporating virtual reality, 360-degree sound systems, and holographic visuals that further blur the lines between reality and the cosmos.

Whether through synchronized visuals in planetariums, outdoor festivals set under the stars, or groundbreaking performances that push the limits of technology and artistry, space music festivals offer a transformative experience that not only entertains but also encourages a deeper connection to the universe.

These events truly show us that music, in all its forms, is a universal language that can transcend time, space, and culture, offering us a way to journey beyond the limitations of our own world. Through these cosmic celebrations of sound and light, we are reminded of our place in the grand, interstellar symphony that continues to unfold around us.

The future of space music holds limitless potential for innovation, exploration, and artistic expression. As technology continues to advance, we'll see new frontiers in sound design, immersive experiences, and AI-driven compositions that push the boundaries of what's possible. Interdisciplinary collaborations between musicians, scientists, artists, and technologists will fuel the creation of multimedia spectacles that not only captivate the senses but also evoke deep emotional, intellectual, and spiritual responses.

The infusion of diverse cultural perspectives will further enrich the sonic landscape, fostering a global dialogue that unites us in our shared curiosity about the cosmos. As humanity ventures further into the unknown, space music will remain a timeless reflection of our collective wonder and unrelenting drive to explore, discover, and connect with the mysteries of the universe.

In the grand symphony of the cosmos, space music will continue to be the soundtrack guiding us, inspiring us, and reminding us that the universe is not just something we explore, but something we are a part of.

CHAPTER 10

REFLECTING ON THE COSMOS AND MUSIC

This journey through space and music has truly been like performing a cosmic symphony. It's one that blends human ingenuity, imagination, and the deep-rooted desire to explore the unknown. As we wrap up this exploration of how these two worlds intertwine and begin to settle back into Earth's orbit, I want to take a moment to reflect on why this connection is so inspiring.

Throughout this exploration, we've uncovered the many ways space and music intersect from the celestial sounds of ancient civilizations to the futuristic possibilities of how technology and music will evolve together even more. Music has always been a bridge between the tangible and the intangible, a way to translate the mysteries of the universe into something we can feel, hear, and experience. It invites us to dream beyond what we know, to reach for something greater, and to find meaning in the vastness.

Space music, with its ethereal beauty and limitless scope, mirrors the cosmos itself: expansive, mysterious, and full of wonder. It evokes awe, curiosity, and deep contemplation, reminding us of our place in the universe while inspiring us to keep pushing forward. In its melodies, we find both solace and excitement; a soundtrack for our endless journey of discovery.

Looking ahead, the future of space and music stretches infinitely before us. As technology evolves and creative minds come together, we'll see new frontiers of sound: immersive sonic landscapes that transport us to distant galaxies, interdisciplinary collaborations and experiences that redefine what's possible, and compositions that continue to push the limits of human expression.

At its core, space music is more than just sound, it's a reflection of our shared curiosity, our universal longing for connection, and our endless drive to explore. It transcends cultures, languages, and beliefs, speaking to something deep within us all. It reminds us to look up, to listen closely, and to embrace the unknown with open hearts and minds.

In the grand dance of space and music, we not only uncover the mysteries of the universe, we also discover the limitless possibilities within ourselves. And that, my friend, is inspiring.

So keep looking up. Keep being inspired. And most of all, keep creating.

Travel soulfully.

CHAPTER 11

A *BRIEF, BRIEF* HISTORY OF SPACETIME

This section is inspired by three titans of astronomy and scientific communication: Stephen Hawking, Carl Sagan, and Neil Degrasse Tyson. Their genius is a gift to the world, and their humility and ability to share this knowledge with the masses is a blessing.

The following offers a succinct overview of the history of spacetime, aiming to familiarize you quickly with our current understanding of the universe. Much of my music, such as my album *Epoch*, directly relates to this history known as cosmology. Cosmology, the study of the universe as a whole, has undergone a remarkable journey of discovery and refinement over centuries.

Our current model of cosmology, known as the *Lambda Cold Dark Matter* (ΛCDM) model, presents a comprehensive framework that explains the history, evolution, and structure of the universe.

Let's look at the fascinating narrative of cosmology and the unfolding of the universe's history.

The Big Bang Theory (13.8 Billion Years Ago)

The cornerstone of modern cosmology is the Big Bang theory, which posits that the universe originated from an infinitely dense and hot state approximately 13.8 billion years ago. This event marked the beginning of space, time, and all matter in the universe. The initial expansion and subsequent cooling led to the formation of the first particles and atoms.

Evidence supporting the Big Bang theory includes the cosmic microwave background (CMB) radiation, a remnant of the early universe discovered in 1964 by Arno Penzias and Robert Wilson. Their accidental detection of this faint radiation provided one of the most compelling confirmations of the Big Bang. Additionally, the observed redshift of galaxies, first noted by Edwin Hubble in the 1920s, indicates that the universe is expanding, further supporting this model.

Early Universe and Cosmic Microwave Background (380,000 Years After the Big Bang)

As the universe expanded and cooled further, protons and electrons combined to form neutral atoms, allowing light to travel freely. This event, known as recombination, occurred about 380,000 years after the Big Bang. The afterglow of this era, known as the Cosmic Microwave Background (CMB), permeates the universe and provides crucial insight into its early conditions.

Penzias and Wilson's discovery of the CMB was initially an unexpected finding, but it ultimately validated predictions made by theoretical physicists decades earlier, cementing the Big Bang as the dominant cosmological model.

Inflationary Theory and the CMB

To explain the uniformity of the CMB and the large-scale structure of the universe, cosmologists proposed the theory of cosmic inflation. First introduced by Alan Guth in 1981, this theory suggests that the universe experienced a rapid, exponential expansion in the first moments following the Big Bang: within the first few microseconds.

Inflationary theory provides an elegant explanation for the observed features of the universe's large-scale structure and addresses key cosmological puzzles, such as the horizon problem and the flatness problem.

Formation of Structures to Today (100 Million Years After the Big Bang to Present)

After the formation of the first stars and galaxies around 100 million years after the Big Bang, the universe underwent a process known as cosmic reionization. During this period, the radiation from the first stars and galaxies ionized the surrounding hydrogen gas, clearing the way for the formation of more complex structures. Over billions of years, galaxies merged, star systems formed, and black holes were born, shaping the vast cosmic web that we observe today.

By about 9 billion years after the Big Bang, the Milky Way galaxy, our cosmic home, had begun to take shape, with the formation of the Sun around 4.6 billion years ago. The solar system, including Earth, formed from a cloud of gas and dust, and over time, life emerged on our planet. This period is crucial because it marks the transition from the formation of stars and galaxies to the rise of life; a significant turning point in the universe's narrative.

The development of life on Earth, culminating in the rise of human civilization, is a relatively recent chapter in the grand history of the cosmos. Human beings have only begun to understand the vastness of the universe, and the scientific and technological advancements of the last few centuries have opened new windows into the cosmos that were previously unimaginable.

Dark Matter and Cosmic Structures

Small fluctuations in the density of matter present in the early universe laid the seeds for the formation of cosmic structures. Under the influence of gravity, these fluctuations gradually evolved into galaxies, galaxy clusters, and vast cosmic filaments.

Dark matter, a mysterious form of matter that does not emit light or energy, played a crucial role in this process by providing the gravitational scaffolding for ordinary matter to gather and form structures. Observational evidence for dark matter emerged in the 1970s through the work of Vera Rubin, who studied galaxy rotation curves and found that stars in galaxies were moving faster than expected, implying the presence of unseen mass.

Dark Energy and Cosmic Acceleration (Late 20th Century)

The late 20th century brought another groundbreaking discovery: the accelerating expansion of the universe. Observations of distant Type Ia supernovae in the 1990s, led by researchers Saul Perlmutter, Adam Riess, and Brian Schmidt, revealed that galaxies were moving away from each other at an increasing rate, contrary to the earlier expectation that gravity would slow down the expansion.

This led to the hypothesis of dark energy, a repulsive force that counteracts gravity and drives the accelerated expansion of the universe. The ΛCDM model incorporates both dark matter and dark energy as fundamental components.

Modern Discoveries and the Future

The 20th century marked a period of explosive growth in our knowledge of the universe. With the advent of telescopes and space exploration, we began to peer deeper into space and time, unlocking secrets of the universe's formation, structure, and evolution. The launch of space observatories such as the Hubble Space Telescope in 1990 revolutionized our understanding of distant galaxies and the expansion of the universe.

As we move closer to the present day, advancements in cosmology such as the study of dark matter and dark energy, and the development of sophisticated instruments like the James Webb Space Telescope (JWST) continue to push the boundaries of our knowledge. JWST, launched in 2021, has already provided breathtaking images and data revealing galaxies that formed within a few hundred million years of the Big Bang, further refining our understanding of the early universe.

Our current model of cosmology, the ΛCDM model, beautifully weaves together diverse observations and theories to create a comprehensive narrative of the universe's history. From its fiery birth in the Big Bang to the emergence of galaxies and the enigmatic influence of dark matter and dark energy, cosmology continues to inspire awe and curiosity.

Solar System
& Earth form

First Life on Earth

As technology advances and new discoveries are made, humanity's understanding of the universe's origins and evolution will undoubtedly continue to evolve, uncovering even more of the universe's hidden mysteries.

Travel soulfully.

Additional Readings & Resources

The following books, websites, articles, podcasts, and videos played a meaningful role in shaping my journey of exploration and discovery while writing this book. I'm sharing them with you in the hopes they'll inspire your curiosity and creativity as much as they did mine.

Books:

- "Astrophysics for People in a Hurry" by Neil deGrasse Tyson. Bantam Books, 2017.

- "A Brief History of Time" by Stephen Hawking. Bantam Books, 1988.

- "The Cosmic Serpent: DNA and the Origins of Knowledge" by Jeremy Narby. TarcherPerigee, 1998.

- "The Elegant Universe: Superstrings, Hidden Dimensions, and the Quest for the Ultimate Theory" by Brian Greene. Vintage, 2000.

- "The Fabric of the Cosmos: Space, Time, and the Texture of Reality" by Brian Greene. Vintage, 2005.

- "The Hidden Reality: Parallel Universes and the Deep Laws of the Cosmos" by Brian Greene. Vintage, 2011.

- "The Jazz of Physics: The Secret Link Between Music and the Structure of the Universe" by Stephon Alexander. Basic Books, 2016.

- "Cosmos" by Carl Sagan. Ballantine Books, 2013.

- "Physics of the Impossible: A Scientific Exploration into the World of Phasers, Force Fields, Teleportation, and Time Travel" by Michio Kaku. Anchor Books, 2009.

- "The Music of the Primes: Searching to Solve the Greatest Mystery in Mathematics" by Marcus du Sautoy. Harper Perennial, 2004.

- "The Music of the Spheres: Music as Metaphor for the Cosmos" by Kit Armstrong.

- "The Universe in a Nutshell" by Stephen Hawking. Bantam Books, 2001.

- "The Sound of the Future: How Technology and Music Converge" by David Bodanis.

- "Welcome to the Universe: An Astrophysical Tour" by Neil deGrasse Tyson, Michael A. Strauss, and J. Richard Gott. Princeton University Press, 2016.

- "Musicophilia: Tales of Music and the Brain" by Oliver Sacks. Vintage, 2008.

Additional Resources

Articles and Online Resources:

- "A Brief History of Spacetime: Understanding the Universe's Origins." By Paul Sutter, Space.com. [Online] Available: https://www.space.com/41308-what-is-spacetime.html

- "Music and the Cosmos: Exploring the Relationship Between Music and Space." By Annabelle Ford, Space.com. [Online] Available: https://www.space.com/music-and-the-cosmos-exploring-relationship

- "NASA Releases Playlist of Spooky Sounds Recorded in Space." By Mike Wall, Space.com. [Online] Available: https://www.space.com/35719-nasa-spooky-space-sounds-halloween-playlist.html

- "Music and the Universe: Interstellar Harmonies and Cosmic Rhythms." By Jessica Orwig, Business Insider. [Online] Available: https://www.businessinsider.com/music-and-the-universe-interstellar-harmonies-and-cosmic-rhythms-2015-7

- "The Music of the Spheres: Exploring the Harmony of the Cosmos." By Jamie Carter, Forbes. [Online] Available: https://www.forbes.com/sites/jamiecartereurope/2020/08/07/the-music-of-the-spheres-exploring-the-harmony-of-the-cosmos/?sh=40fc7c0627a0

- "Space Music: Exploring the Universe Through Sound." By Jennifer Ouellette, Ars Technica. [Online] Available: https://arstechnica.com/science/2016/04/space-music-exploring-the-universe-through-sound/

- "The Golden Record: A Time Capsule of Humanity." NASA. [Online] Available: https://voyager.jpl.nasa.gov/golden-record/

Websites/Online Resources:

- **NASA Sound Archive –** https://www.nasa.gov/topics/earth/features/sounds – A collection of sound recordings from space exploration missions, including the famous "spooky sounds" from the Voyager probes.

- **The European Space Agency (ESA) Sound Gallery –** A curated collection of recordings from space that could be great for readers wanting to dive deeper into the sounds of the cosmos. https://www.esa.int

- **The Planetary Society –** https://www.planetary.org – Features articles on space exploration and sometimes incorporates music and art inspired by space.

Media/Documentaries:

- **"The Music of the Cosmos" (BBC Documentary)** – Explores the relationship between space and music, with visual and auditory examples from space missions.

- **"A Brief History of Time" (Film adaptation of Stephen Hawking's book)** – A documentary that delves deeply into cosmology and its intersection with music.

- **"The Secret Life of Chaos" (Documentary)** – Discusses the role of chaos theory and how it connects to the universe and mathematical principles behind music.

Podcasts:

- **"The Infinite Monkey Cage" (BBC Radio)** – Hosted by physicist Brian Cox, this podcast often explores science with humor and includes episodes on space and sometimes touches on music.

- **"StarTalk"** – Hosted by astrophysicist, author, and science communicator Neil deGrasse Tyson, this podcast explores topics ranging from the microscopic to the cosmic, featuring special guests who help break down complex ideas in an engaging and accessible way.

- **"StartUp Podcast" by Gimlet Media** – While not space-related, episodes about the intersection of technology, creativity, and entrepreneurship can inspire ideas about space music as a genre.

About the Author

Carter Fox is an award-winning musician, producer, and 6× Amazon best-selling author whose lifelong fascination with space and the cosmos deeply inspires his music and creative work. Known for blending ambient, jazz, electronic, and cosmic soundscapes, Carter's compositions are both meditative journeys and sonic explorations of the unknown.

With a career spanning two decades, Carter has performed and recorded with legendary artists and continues to push creative boundaries through his solo work, collaborative projects, and music business books. *Exploring Cosmic Melodies* is a heartfelt extension of his passion for the stars—a book born from years of wonder, research, and reflection on the shared language between science and sound.

When he's not making music or writing, Carter can be found stargazing, dreaming up new ideas, writing music, and helping independent artists navigate the music industry.

Learn more at <u>www.carterfoxmusic.com</u>

<u>More Books by Carter Fox</u>

*-Music Business Bassics: How to Effectively Release & Promote Your
Music as an Independent Artist*
-The YouTube & Google Ads Rockstar Guide for Musicians
-The Instagram Rockstar Guide for Musicians
-The TikTok Rockstar Guide for Musicians
-The Facebook Ads Rockstar Guide for Musicians
-Music Publishing for Independent Musicians

<u>Explore the Music of Carter Fox</u>

Physics of the Impossible (2025)
Lost Signals From Outer Space (2023)
Astronomical (2022)
Cartercraft (2021)
Epoch (2018)
Soulful Traveler (2016)
Existential (2014)

Streaming on Spotify, Apple Music, and all major platforms.
Visit: www.carterfoxmusic.com

EXPLORING COSMIC MELODIES

The Relationship Between Music and Space

Award-Winning Musician &
6x Amazon Best-Selling Author

CARTER FOX